Martha Schwartz Partners Landscape Art and Urbanism

景观艺术与城市设计
玛莎·施瓦茨及合伙人设计事务所作品集

[德] 马库斯·詹斯奇（Markus Jatsch） 主编
杨至德 译

江苏凤凰科学技术出版社

景观无所不能，景观无所不有。
玛莎·施瓦茨

Landscape can

be about anything

Martha Schwartz

序 / Foreword

20 世纪 80 年代，美国景观建筑学在生态学的束缚和控制中缓慢产生。此前，生态学曾严重打击了景观设计作为创造型行业的倾向。尽管科学和艺术是组成学科和专业的两大基本元素，但在数十年的发展中，科学占了上风。在很大程度上，受伊恩·麦克哈格划时代的著作《设计结合自然》的影响（1969 年第一次出版发行），景观设计被赋予"规划多于发明，分析多于创造"的特征。20 世纪 80 年代中期，当玛莎·施瓦茨进入景观设计行业时，她的作品就像一枚空中炸弹，在那些仍然遵循弗雷德里克·劳·奥姆斯特德的传统并采用"自然"方式排列的绿色植被的上空爆炸了。

或许是因为她的艺术学而非园艺学背景，或许是因为她特立独行的性格，施瓦茨最早发表的作品"面包圈花园"——首先是她的想法——打破了景观设计业内的"自满"。大量反对的声音蜂拥而至，质疑施瓦茨的这个小作品的合理性："这（例如，线性排列的面包圈）不是景观设计。"但之后不久，那些反对者不得不停下来认真思考："那么，景观设计究竟是什么？" 无论如何，由她所引起的对景观设计实践的重新思考，为美国景观设计行业作出了巨大贡献，其影响甚至远及国外。

从一开始，施瓦茨的设计实践就横跨在艺术和景观设计之间的轴线上，而非同时展现它们的特质——因为在某种程度上，它们是相互矛盾的。在大多数情况下，进度安排、项目场地以及气候条件可以定义一个"设计"项目。但是，得益于她对形态和空间所持有的不拘一格的思想、对自然主义的反对、对资源的精心考察以及对材料的收集与使用，施瓦茨的设计实践更接近于艺术。按她自己的话来说就是："（艺术和景观设计）

American landscape architecture in the 1980s was only hesitantly emerging from a grasp by ecology that had seriously dampened aspirations for landscape design as a creative enterprise. While science and art comprise the two components of the discipline and profession, for over a decade science had prevailed. Due in large part to the influence of Ian McHarg's epochal book *Design with Nature*, first published in 1969, landscape practice privileged planning over invention, and analysis over creativity. When Martha Schwartz entered the scene around the middle of the decade, her work landed like an aerial bomb that blasted the greenery still being planted in the "naturalistic" clumps so dear to those following in the tradition of Frederick Law Olmsted.

It may have been her background in art rather than horticulture, or perhaps her wayward personality, but from the earliest published project—a garden limned with bagels, of all things—Schwartz's ideas drove a wedge into the profession's complacency. Against her personal claims of legitimacy for this small work, a flood of retorts asserted that: "This (i.e. bagels in a line) was not landscape architecture." But perhaps soon thereafter, each of the disclaimers was forced to pause for a moment and consider: "Well then, just what is landscape architecture?" At the very least, her instigating any reconsiderations of practice constituted a significant contribution to landscape architecture in the United States and, to some degree, even abroad.

From the start Schwartz's practice has straddled the fine line between art and environmental design—not that the two approaches should be, in any way, incompatible. In most cases, of course, an address of program, site and climate immediately qualifies the project as "design". But a certain errant attitude toward form and space, her rejection of naturalism, and a resourceful investigation and use of materials, nudge the meter closer to art practice. In Schwartz's own mind, the two were neither independent nor oppositional.

序

这两个方面,既不是相互独立的,也不是相互对立的。"

20世纪80年代的景观设计作品大多被"即时开发"所影响而成为相应的"即时景观设计",其中,商业建筑设计多于民居设计。佐治亚州亚特兰里约购物中心中的金色青蛙,运用美国本土景观元素,并通过网格布局——一种结合了极简艺术学与现代建筑学特征的"艺术排序",达到一个更高的审美境界。

通过运用镜像凝视球,美国庭院中一些常见的景观元素——被建筑设计师罗伯特·文丘里称为"丑陋且普通的元素",在新的环境和结构之中蜕变为优雅且脱俗的景观。然而,批评再一次指向了施瓦茨设计所体现的嬉闹和"反自然",他们仍然声称那不是景观设计。然而,随着时间的推移,这种批评渐渐减少,最后几乎消失。

从更深层次的角度,施瓦茨与其早期的合作伙伴——第一个是彼得·沃克,后来是肯·史密斯和大卫·迈耶——密切交流,相互影响。施瓦茨-沃克所营造的景观严谨有序,由网格结构和大片线条组成,充满活力且极具嬉闹感。例如,加利福尼亚州圣地亚哥市滨海线性公园,把铁路路基、主干街道和人行道编织成一块巨大的"塞拉普地毯"图案——这实际上是把原有的基础设施重新组合成一系列颇具活力的设计元素。人们会说,施瓦茨与肯·史密斯的交流强化了设计创作中的流行色彩,而大卫·迈耶则掌控全局,致力于实际项目的材料和细节。他们在多伦多约克维尔公园项目中的合作充分展示了其同时表达艺术、布局和人类行为的能力。后来,施瓦茨离开了她的合作伙伴,在马萨诸塞州剑桥创立了自己的第一个工作室,并于十年前搬到了伦敦。

看看这本令人印象深刻的作品集,从这些已经建成的项目中,我们可以感受到施瓦茨对合成材料和生物材料的喜爱。事实上,施瓦茨并不依赖于某种单一的材料或者媒介,她更看重掩藏在材料或媒介之中的设计理念及最终的建成效果。每种介质都有其价值和效果。然而,像拥有很深厚的园艺功底的景观设计师的作品那样,以某种性质特殊的植物作为设计基础,这种情况确实非常少见。取而代之的是像陀思妥耶夫斯基那样——施瓦茨相信"量刑应该与罪行相当"。什么是合理?什么是有效?什么是感觉正确?施瓦茨运用植物的最著名案例,或许是位于德国慕尼黑的瑞士再保险总部大楼周边景观。在这里,植物与矿物质材料相结合,构成一系列主题花园,为建筑四周铺上一层"地毯"。每一面都采用色彩各异的植物和岩石——藤蔓沿着建筑表面一直爬行到顶端,使建筑与景观融为一体。每一个条带都由单一的植物或惰性矿物质材料(比如,彩色碎玻璃或砾石)组成;在不同的季节,不同的立面尽显不同的美丽。例如,红色的一面,正如办公室工作人员所描述的:"秋天是卫矛叶子所营造的一片红色火焰,春天是球茎类植物闪光的红色花朵,冬天是灌木上挂满的红色浆果。"在更加依赖于合成材料的景观之中,最典型的是2000年在日本岐阜打造的北方町屋景观,在那里,有机玻璃发挥了重要作用。在该项目中,所有社区设计都出自女性之手,景观设计也如此。社区中央的系列凉亭巧妙地运用彩色亚克力材质,把美妙的色彩投射到

Foreword

The 1980s work was colored by a plethora of instant landscapes that accompanied instant developments, at times for housing, more often for commerce. The golden frogs in the Rio Shopping Center in Atlanta, Georgia, drew upon the elements of the American vernacular landscape but elevated them to higher aesthetic provinces through their arrangement in a grid—an ordering that was so much a hallmark of minimalism in art and modernism in architecture.

Using mirrored gazing balls and other common elements from the American backyard, the ugly and ordinary—as described by the architect Robert Venturi—became the elegant and extraordinary in their new contexts and structure. Again, criticism attacked the playfulness and outré nature of Schwartz's work. The claims that this was not landscape architecture. Over time, however, the arguments weakened considerably and in time almost disappeared.

Seen from a position beyond the drafting room, the early partnerships—first with Peter Walker, and later Ken Smith and David Meyer—were mutually influential and times of fecund exchange. The Schwartz-Walker landscapes were strictly ordered, often structured by grids and fields of lines yet energized by a certain playfulness. The Linear Marina Park in San Diego, California, joined the railroad beds, avenue, and pedestrian ways within a greater "serape" pattern—in effect, reconfiguring the infrastructure as a vital element of the design. The exchanges with Ken Smith, one might suggest, reinforced the pop nature of the office's production, with David Meyer keeping things in check and contributing to the materiality and detailing of the realized projects. Their collaboration on the Yorkville Park in Toronto demonstrated a capability for addressing—simultaneously—art, locale, and behavior. Schwartz left those associations behind and moved on, opening her own office, first in Cambridge, Massachusetts, and lastly in London.

Reviewing the impressive portfolio of realized projects one observes a fascination with synthetic as well as living materials. One would venture, however, that to Schwartz the question is not a reliance on any single material or medium, but instead a concentration on the idea behind the design and the ultimate perception of the landscape as realized. Each medium has its values, each its effect. But only rarely, if ever, has any design been based on plants of some characteristic species, as works by landscape architects deeply rooted in horticulture might. Instead, like Dostoevsky, she believes that "punishment should fit the crime." What is appropriate? What is most effective? What feels right? Perhaps Martha Schwartz Partner (MSP)'s most brilliant use of plants is found at the landscape that surrounds—and mounts—the Swiss Re offices in Munich, Germany. Here the design intermixes plants and minerals in a series of themed gardens that carpet the understory of the building on four sides, each articulated with varied colors in planting and rocks—culminating in the vines that climb the façades of the buildings and literally root the architecture in the landscape. Each stripe is made up of a single plant species or inert mineral such as crushed colored glass or gravel. Through the seasons each of the quadrants rises to prominence. In the red quadrant, as the office describes it: "the Euonymus elates is a blaze of red in autumn, a field of bulbs blooms with a dazzling red in the spring, and shrubs provide red berries in winter." Of the landscapes relying more completely on synthetic materials, Plexiglas appeared to great effect in the landscape for the Kitagata housing estate in Gifu, Japan, completed in 2000. All the housing blocks were designed by women, as well as the landscape. The central pavilions sheathed in polychrome acrylic lent splashes of nuanced color

| 序 | Foreword | 6 |

银灰色的住宅单元上，同时把耀眼的阳光过滤成彩色的光斑，为在凉亭中休息和娱乐的人们提供一番享受。尽管在该项目中，合成材料扮演着主要角色，但生物材料依然是不可或缺的设计元素——它们为设计增添了色彩和生机。否则，那里就只是一整片灰色建筑。

在她的职业生涯中，施瓦茨在三个"前线"上坚持不懈地战斗着。第一个，是她作为艺术家和景观设计师所付出的努力。在一个逐渐被科学、分析和学术研究主导的行业中，她始终坚持"设计植根于创造之中"。作为教育工作者，她重视过程和结果，然而她更强调在现实生活中人们情绪和心理上的最终感受。同时，发明、创造以及不同形式、色彩、材料之间的转换，始终是她最大的兴趣所在。

其次，是她作为一名女性的身份和立场。在20世纪初期的景观设计行业中，曾有许多知名的女性活跃在这个行业并从事重要工作。然而在战争期间，她们的人数有某种程度的减少。在随后的几十年中，景观设计工作室从单独从业或合伙从业的形式转变为公司结构的形式。在这种情况下，缺乏个性的女性逐渐淡出。毫无疑问，在美国有许多杰出的项目是以女性为中心的。但是，在一些大型公司中，女性主要作为合作人或合伙人。然而，几乎没有其他女性能够达到施瓦茨在业内的高度，尽管她们有可能完成过更多的景观设计项目。施瓦茨为留任哈佛大学所做的斗争表明，在景观设计行业，一些残存的封建思想仍然影响着广大女性。这也进一步证明，施瓦茨对既有成规发起挑战的决心和愿望。这或许就是她在中国及其他国家开展项目取得成功的原因之一。她永远是一位"破坏者"，起初是"儿童捣乱分子"，现在是真正的"捣乱分子"，正如法语中所说的"正当年"。

随着时间的推移，景观设计在方法、技术和规模上都不断成熟。在某种程度上，早期作品更倾向于艺术，凭借直觉把设计理念应用于场地中，具有一定的审美独立性。相反，后期作品更加注重场地本身，在现有条件的基础上打造新景观，并与甲方的设计要求相符合。20世纪90年代，在玛莎·施瓦茨及合伙人设计事务所的诸多项目中，英格兰曼彻斯特交易广场项目充分展现了景观设计对已衰落地区的振兴能力——该项目场地曾因恐怖主义活动而被污染。然而，在其他地方，应该说是在其他案例中，我们得到的就是一大片平地——虽然不是一无所有，但我们还是被告知了场地的土壤、地形和气候信息，除此之外，几乎没有可供景观设计利用的突出元素。在这种情况下，（对景观设计来说）一个正式的设计声明更加有效，建筑也更加重要。在某些城市设计项目中，绿地与地面铺装结合在一起——这在高层建筑环境中能够起到某种缓和作用，尽管在水泥地块上建造景观受到些许限制。在明尼苏达州明尼阿波利斯联邦法院大楼前面的广场上，我们引用了当地特有的景观元素，以当地树种的树干作为座椅，并堆成沙丘状，以象征明尼苏达州冰川地貌中的鼓丘。乍一看，这个设计更偏向于建筑领域，更适用于工程领域，更加优雅，更加依赖于几何学。然而，它却是造型特殊且具有地域色彩的。因此，施瓦茨认为，巧妙利用自然或建筑元素的变形并将其转换为新的形式的是至关重要的，而不可单纯地复制"自然"景观。

to the grey and silver housing units and filtered brilliantly tinted light to those seeking refuge and pleasure within the pavilions. While synthetic products played a key role, living materials completed the palette that added color and life to an otherwise gray architectural ensemble.

Throughout her career Schwartz has continued to simultaneously wage war on at least three fronts. First, was her campaign to operate as an artist as well as landscape architect. In a profession increasingly dominated by science, analysis, and academic research she has staunchly maintained a position rooted in creative response. As an educator and lecturer, she has considered both process and product but has stressed the importance of what we ultimately experience in actuality, emotionally and psychologically. The invention, manipulation, and play with forms, color and materials, has remained a central interest.

Then there was her identity and stance as a woman. In the early decades of the twentieth century, a number of notable women maintained active and significant practices in landscape architecture, although their number to some degree dwindled during the war years. In the decades that followed, as the organization of landscape offices turned from the sole practitioner or partnership to those more corporately structured, the individuality, if not presence, of women diminished. Without doubt, in the United States several outstanding practices are centered on women, while other women serve as associates or partners in larger offices. However, few (if any of them) have maintained an equally high profile as Martha Schwartz, although they may have actually realized more landscapes. Schwartz's battles securing tenure at Harvard also illustrate the residual attitudes still affecting women in landscape architecture, and testify to her resolve and willingness to tackle the establishment when required. This may be one of the reasons she has been successful in executing projects in China and other countries. She has also remained the eternal provocateur, first as an enfant terrible, and now as a terrible, as the French say, "of certain older age".

Over time the approach, sophistication, and scale of the commissions have all increased. In some ways the earliest work, more art oriented, applied ideas to the site in a gut response executed with a certain aesthetic independence. Later works, in contrast, have drawn more from the site, transforming the pre-existing conditions into a new landscape that responds to the stipulations of the design brief. Maturity has set in. MSP projects of the 1990s, like Exchange Square in Manchester, England, demonstrate the ability of landscape architecture to reinvigorate a site fallen into decline, in this case one tainted by an act of terrorism. Yet in other places, often in other lands, the site lay as a vast plane—not exactly a tabula rasa, of course, given the existing soil, topography, and climate, but with few outstanding features upon which to hang a design. In these instances a formal statement becomes more valid and the structure more prominent. In some urban projects a matrix of greenery and paving may join—and to some degree soften—a set of high-rises despite the limitations of a landscape constructed on a concrete slab. The plaza fronting the courthouse in Minneapolis, Minnesota, referred more specifically to the state's landscape, using the trunks of native trees as seating and mound-forms suggestive of the drumlins of Minnesota's glacial landscape. In the first instance, the approach was more architectural, more applied to the world of construction, more polite and more reliant on geometry. The allusions were specific and local, The transformation of natural or architectural references into new forms is key; there has never been any attempt to replicate a "natural" landscape.

Foreword

The dimensions of many projects have grown exponentially over time. When one compares tiny works like the early Splice or Bagel gardens with expansive waterfront schemes and housing developments for hundreds of thousands of inhabitants, the nature of the differences and the differences in their natures become evident. One now witnesses in the designs a greater acknowledgment and understanding of the panoply of conditions that shape a landscape and the knowledge necessary to address the full range of design considerations. While the work of MSP today always maintains a noticeable identity. The complex network of factors that comprise today's landscape practice enriches rather than diminishes the resulting landscape. Consultants are required; the process is more complicated and more collaborative. While on some projects the architecture seems to lead, the landscape integrates buildings with the greater habitable environment of which the buildings are only a part. It is no wonder that many, if not most, of these larger works are located in China. The population is immense; the economy offers the financial means; the lands is available, as is the will to look toward the future rather than return to the past. These conditions have provided a situation open to creativity and innovation, in form and space as well as method. Given the vast differences in scale, economic resources, location, and time frame the question is: How can a landscape architect maintain a consistent stance from art installations to town plans?

To me, the answer has been a consistent use of abstraction. Abstraction, because of its simplification or distance from specific reference, is often considered as a quality divorced from reality. This is not the case, at least not historically. The origin of the word derives from the Latin term "to draw from". One begins with the conditions of a particular situation and from it one draws what is key, what is pertinent, what needs to be reformed, reshaped, or at times intensified. For example, one may abstract topography or vegetal forms as a basis for patterns and order. From the existing conditions one devises an approach, albeit incorporating an existing design vocabulary that is personal and particular. One draws from and applies to, and by this manner Schwartz's design vocabulary, while to some degree consistent, has continued to evolve—at times applying a choppy biomorphic line we might call "bio-cubic", or smooth curves that suggest flows; or forms more staid where the situation, normally urban, calls for restraint.

Quite unusually for most landscape architects, Schwartz's practice still embraces both the macro and micro scales, with projects that range from large scale site planning to the diminutive art installation, several of which have been collaborations with Allison Dailey. Outstanding among these is "City and Nature", a perceptual labyrinth for a garden show in Xi'an China which manipulated physical passage with visual enigmas that result from periodic shifts in transparency and reflection. The materials were vernacular and timeless: grey brick, willows, one-way mirrors, and bronze bells. If the materials were common, the configuration was radical. Here the artist also served as perceptual psychologist and social scientist, not to mention a wizard who conjured surprise and delight. Where public art is often an alien artifact inserted into a place, landscapes such as these become new things characterized by an inextricable link with the site and insights into human behavior often lacking in work by artists less familiar with public space.

Martha Schwartz has noted that when she entered graduate landscape studies at Harvard she was told that art and land-

| 序 | Foreword |

过一样；生态学本身塑造自身形态，但缺乏美感；景观设计师或艺术家，必须在景观和艺术之间作出选择，不可二者兼而有之。然而，近30年来（用她自己的话来说）"吉卜赛"式（办公地点经常变化）的实践，施瓦茨与她的合作伙伴一直勇敢地面对并不断否定这些清规戒律。其作品经常具有某种标志性形态：一种并非自然、原生态的形态。植根于生态学中的景观，仍然可以彰显出美学特征。新建景观必须充分考虑自然过程，但并不必刻意模仿自然形态。实际上，对景观的设计和改造可以改善或美化一个令人头疼的地表形态。最重要的是，"良好"的景观在支持自然和人类存在与发展的同时，也为人们带来惊奇和欢乐。施瓦茨甚至勇敢地提出，作为为之奋斗的目标与理想，她将一直追求那令人难以理解的美和更多的嘲笑！

马克·特莱伯
美国加利福尼亚大学伯克利分校建筑学荣誉教授

scape do not mix; that "good" landscape should appear natural, as if untouched by human hands; that ecology produces its own forms and lacks an aesthetic; that one must choose: be an artist or make viable landscapes as a landscape architect. You can't have both. Through thirty years of practice, she and her collaborators—in what she called her "gypsy" practice, with the location of its offices shifting over time—have confronted and all but negated those precepts. The work always possesses an identifiable form; it is not nature as it was. Landscapes rooted in ecology can have an aesthetic; a new landscape must understand natural processes but has no requirement to mimic natural forms. That design and its realization can actually ameliorate a troubled situation. And most of all, that while supporting both natural and human existence, "good" landscape can also delight. And she has even dared to propose, as a desirable goal, the pursuit of that elusive and much derided word, beauty.

Marc Treib
Professor of Architecture Emeritus
University of California, Berkeley

| 前言 | Introduction |

这本书的编写与出版对于介绍并展示我们多年来的设计实践意义非凡。书中收录了我们从1979年开始一直到最近30多年的大部分作品。在这对我而言意义非凡的30多年中，景观建筑学本身经历了令我意想不到的质变，并且，这种变化正伴随着全球现实持续着。

鉴于建筑学被世人誉为"变化的催化剂"，我被带向了跌宕起伏的漂流之旅，在一条需要不断适应新变化的轨道上持续前行。通过景观设计实践，我了解了这个充满神奇的世界，并使工作室的形式与之相匹配——就像蒲公英的种子随风飘动——工作室几经变迁。为了适应不同的时间和地点，工作室几经重塑。波士顿是我的第一站，在那里我的第一个作品诞生了——这个手工艺术装置挑战了传统景观建筑学的惯例。之后我来到纽约，接受阿奎建筑设计公司的委托，这是我们第一个"真正"的项目。在纽约之后，我们转移到旧金山，继续从事一些容易预见效果的小型项目，与一些预算紧张却仍期望有趣设计的开发商合作。从加利福尼亚开始，我们赢得了一个日本项目——那是第一个国际性项目的委托。

这种"吉卜赛"式的实践一直持续到1992年，之后我又回到马萨诸塞州剑桥，在哈佛大学设计学院任教。至今，我仍然在那里教学。从这个时期开始，我们开始接触很多美国本土和欧洲的公共项目。特别是欧洲，那里的人们对"公共区域"概念更感兴趣，而那时的美国人还不太接受这个概念和相关委托。由于1996年在曼彻斯特广场项目中的成功以及欧洲市场对于公共区域价值较高的接受程度，我们于2005年在伦敦设立了一个全新的工作室。"迁居"至伦敦，意味着我们更加接近欧洲市场——欧洲城市的市长们意识到公共景观在保持城市竞争力方面的价值。

This is an important book for our practice. It represents a compendium of work that spans from 1979 to our most recent projects: a period of over thirty years. During what has been a significant period for me, the profession of landscape architecture has also undergone a metamorphosis that I could never anticipated and one which continues to transform in response to global realities.

As landscape architecture engages with this world as a place of accelerating change, my activities have been propelled on a continuous journey into unchartered waters and on a trajectory that constantly requires adaptation and new responses. Through my landscape practice, I engaged with this world of wonder and adapted to its alterations with a studio model that, like a dandelion seed, floats along the currents of the wind. In this sense, our office has travelled from place-to-place, touching down as we re-established ourselves at various points in time and in consecutive locales. Boston was the first location founded upon my earliest work; building art installations as manifestos that challenged traditional landscape architectural conventions. New York followed, where the first commission by Arquitectonica determined the focus of work for a "real" project. In San Francisco, subsequently, we continued to do small but highly imaginable landscapes, working with developers on extremely tight budgets, who, nonetheless, wanted to do interesting work. It was in California that our first international commission was won and work in Japan opened up to us.

This kind of "gypsy" practice continued when I returned to Cambridge, Massachusetts, in 1992 to begin teaching at the Harvard Graduate School of Design, where I still teach today. It was at this point we also became more involved with public work in the USA, as well as in Europe, because they were demonstrating more interest in the existence of a "public ream". This was a concept and a commitment that America had yet to embrace.

| 前言 | Introduction |

此外，对公共区域的艺术或景观设计的投资，在很大意义上就是对城市文化的投资，并可充分展示城市文化的包容性。同样令人兴奋和鼓舞的是，在环境保护理论方面，欧洲比美国领先许多，当欧洲许多城市开始向"健康"城市转变时，美国的一些城市才刚开始对屋顶绿化感兴趣。在这种环保意识较为高涨且城市历史氛围较为浓郁的环境中工作和学习，对我来说宛如置身天堂。另外，伦敦至欧洲和中东地区几乎相等的距离和它当时即将成为国际金融中心的事实，使它吸引了很多顶级专业人员和海外客户，这些都使我们获益良多。实际上，2004年至2009年，我们在这两个地区做了大量工作。

现在我们发现，在中国的工作面临着各种挑战，但设计实践却异常精彩且丰富。我们"遭遇"了中国当下如火如荼的改革和反乌托邦的倾向。然而，中国人学习新事物的速度非常快，他们将在不久的将来成为气候变化和环境立法领域的重要领导者。中国人乐意接受新思想且不惧怕变革，这使中国成了一个文化氛围浓郁而充满雄心壮志的地方。面对如此庞大的中国市场，作为风景园林设计师，把环境友好型和步行优先型的城市规划思想引入正在发展中的社区，正是天时、地利、人和的好机会。为了配合在中国的工作，我们在上海设立了一个小型办事处，帮助理解中国文化并协助中国项目的实施。因此，我们仍以"吉卜赛"的方式在世界各地新兴地区漫游，为那些需要我们的客户设计具有地标性、可持续性、多层次价值且被大众接受、与大众互动的景观。

我的专业背景始于我对景观专业的一无所知，在艺术学校度过了童年时代并完成了大学本科学习之后，我终于在1974年无意选中景观设计作为继续深造的专业。当时，我对这个专业几乎毫不了解。我选择景观设计专业的原因并不复杂：我想要学习如何创作大型艺术作品。那时的我是一名"大地作品"艺术家们的追随者——"大地作品"曾是20世纪六七十年代全球艺术界的一个焦点。一些艺术家（比如，迈克尔·海泽、理查德·朗、沃尔特·德·玛丽亚和安德烈·卡尔）的作品走出了艺术展馆，伫立于美国东南部的景观之中。这些"史诗"般的作品与其周边景观完美结合并交相辉映。这些艺术家堪称新兴环保运动的先驱，让人们意识到景观的美丽，并通过全新且富有现代感的镜头欣赏一个个美丽的景观。这也是"场地专化艺术"第一次进入我们的设计语汇之中。我希望我的艺术作品与特殊的场地进行互动，并在城市环境中进行艺术探索。在那时，尽管我认为加入景观设计行业是一个帮助进行艺术探索和实现理想的合理方法，但很快我发现，在全班30人之中，我是仅有的两名拥有纯艺术专业背景的学生之一。在研究生院学习的第一年，我被灌输了五个重要思想，它们深刻地影响了后来我在业内的定位：

（1）"好"的景观，是不体现人为雕琢痕迹的景观；
（2）艺术与景观之间没有关联；
（3）环境议程与艺术创作之间没有关联，只能二选一；
（4）如果你成了建筑的一部分，那么你就成了问题的一部分；
（5）生态学中的确存在美学。

Thus, due to the popularity of our plaza in Manchester in 1996, we were drawn to Europe where there was greater receptivity to the value of public space and we were able to locate a fledgling office in London 2005. The move to London meant we were closer to the European community, where mayors were aware of the value that public realm landscape contributed to keeping their cities competitive. Furthermore, investment in public space was very much part of an urban culture that accepted design and art in the landscape as fully compatible. Equally exciting and inspiring was the fact that the Europeans, in comparison to the US, were far more advanced in environmentalism, to the point where they had moved to the scale of "healthy" cities whereas in the US they had just begun to become interested in green roofs. This conducive atmosphere provided opportunities to work and learn in a more environmentally aware culture that also had a long urban history: it was like being in heaven for me. The proximity of London to Europe and the Middle East, plus the fact that London was becoming an international financial hub that drew top professionals and clients to propose work on foreign soils also benefited us. As a result, we worked heavily in Europe and the Middle East from 2004 to 2009.

We now find ourselves working prolifically in China which has many challenges and affords many new stimulating opportunities. We encounter the transformation that China is undergoing at every turn and some of the dystopic tendencies as well. Yet China produces such rapid learners that they will soon be a source of global leadership in climate change and environmental legislation. Their people are open to new ideas and unafraid of change making China a place of cultural and youthful vitality as well as ambition. As landscape architects, in this large new market, there is a phenomenal opportunity for us to help bring more environmentally friendly, pedestrian-based city planning into the awareness of the development community, making it a great time and place for the profession of landscape architecture. To complement our work in China, we have a small office in Shanghai that helps us to translate the Chinese culture and to get our work built. So, our gypsy practice continues to wander around the globe into new areas across the world to serve those who call on us to do what we do: design iconic and sustainable landscapes that people love, interact with and that will create value at many levels.

My personal background started with knowing nothing about the profession when I randomly decided to attend graduate school in landscape architecture in 1974 after spending my childhood and undergraduate years in art schools. My reason for choosing landscape architecture was unsophisticated except that I knew I wanted to learn how to build big art. I was an avid follower of the "Earth Works" artists who came into the art-world spotlight in the late 1960s and early 1970s. I was enamoured with artists such as Michael Heizer, Richard Long, Walter de Maria and Carl Andre who created works that went outside the gallery and were built in landscapes of America's southwest. They were heroic works integrated and resonating with the landscapes in which they were situated. They were also bell-weathers of the exploding environmental movement making us aware of the beauty of these landscapes by allowing us to see it through a new and contemporary lens. This was also the first time that "site specific art" entered our vocabulary. I knew I wanted to do art that interacted with a specific site and use the urban context for artistic exploration. I considered that entering landscape architecture was a reasonable way to explore these ideas but I quickly surmised that myself and one other person were the only two with a background in fine arts in our class of thirty. In that

Introduction

first year of graduate school, I was taught five important things that deeply influenced my subsequent position in the field:
(1) A "good" landscape was a landscape that did not show the hand of man;
(2) There is no connection between art and landscape;
(3) There is no connection between having an environmental agenda and making art: one had to choose between the two;
(4) If you are part of building, you are also part of the problem;
(5) Ecology does have an aesthetic.

I did not agree with any of these "important" statements, in fact I have spent the last thirty plus years using these preconceived ideas about what a landscape "should" be as a spring-board for the work that we do in our practice. I have approached the field from an artist's stance and to question, and subvert the status-quo to arrive at new propositions. I see the landscape as an artist's medium, with a set of materials to work with; earth, water, sky and living plants, as well as any other materials that are necessary for self-expression. With these materials and a good imagination, landscape architecture can be a cultural art form like sculpture, painting, dance or architecture.

An environmental ethos that underpins any landscape work and makes it a function is included within this idea of the artist's medium as well that a built landscape must engage with and embrace people as part of its role: it is not a static device. Our living behaviour, psychology, culture and society must be included as part of any comprehensive idea of urban ecology if we are to produce sustainability. This idea is reflected in the dialectic between man (bad) and nature (good) as an old one; but when played out within the realm of "landscape" it becomes even more fuelled by our own distinct images and mythologies held by individual cultures, many of which clash fiercely when confronted with the facts of our rapid urbanization. Without clarity that we construct the landscapes and the nature in which we live, that they are produced, then landscape will continue to be trivialized to perform as romantic remnants of a beloved image about which we fantasize but in reality has no resemblance to the actual environments in which we work and live. These fantasies of nature, that we carry close to us, like a beloved teddy bear, prevent us from acting realistically and strategically. This only deepens the crisis in which we appear powerless to stop the environmental and visual degradation that has become our world in the 21st century.

The importance of this environmental effort is tandem with the realization that we all live within limited resources and the profession has expanded greatly within the last 30 years in response. We are the green profession, par excellence, so our voices are louder and our skill sets are needed now more than ever. We are making progress in helping people to understand that we must go beyond the garden to understand the urban landscape as a functioning and multi-layered system that underpins the building of cities. Human health and a good quality of life for people results from these healthy and environmentally functional landscapes especially as cities densify that allow for more efficiencies of resources making the man-made islands (cities) more desirable.

Living in cities also allows for more effciencies of resources, so the conundrum exists: how do we assist people living on dense, man-made islands (cities) to live in a way that is most environmentally friendly, low-carbon and creates a high quality of life? The answer is to build cities that are based on sound ecological planning and to design these cities for people. Without putting these two goals foremost and together, we can

前言 | Introduction

噪声和污染所压倒。邻里社区之间或与其他城区之间，通过林荫人行道和自行车道彼此串联，进而与更大范围的城市结构和基础设施连为一体。此外，对于提高人们的生活质量而言，在多种类型的绿地和开放空间之中，在紧张的城市生活之中，放松和娱乐身心是非常重要的。最后，人们总是试图在生活中发现美。所有这些特点都有助于营造令人心驰神往且健康宜居的城市。我们最大的希望是，以设计精良的景观主导型总体规划营造可持续性城市环境。

加勒特·艾柯博于1950年编写出版了《人性化景观》一书，他重点论述了景观创造性与社会互动性之间的关系。现如今，这个话题在景观建筑领域中很少被谈到。我们希望景观设计超越场地范围并扩大到城市范围的同时，似乎忘记了人类尺度，以及通过艺术和设计创造的人与人情感连接的价值。加勒特·艾柯博和劳伦斯·哈普林，两位现代景观设计领域的先驱，都是人道主义者。他们认为，景观设计应该把关注的焦点放在人类本身，同时与自然和谐相处，致力于打造令人喜爱和欢迎并具有文化重要性的空间场所。随着行业范围的扩展，业内展开了关于"尺度"的讨论。在理论上，我们应该打造各种尺度的景观。然而，一方面，我们必须在大尺度范围内运用城市主义设计理念，把各种信息加以整合，提出多层次的规划战略，以解决复杂的城市问题；另一方面，我们也不能忽视人性化尺度在实体设计上的重要性——相反，这将使我们获得尊重和认可。所有设计细部都是为了向客户提供高质量的产品。在人性化的尺度上，人们可以知道，这块场地在设计方面是否怀有敬意，是否具有幽默感和个性。一个设计讲述一段故事，形成关系链接，为那些置身于空间之中的人们创造价值；一个设计如果不考虑人的价值，那么就无法实现可持续发展。

高质量的设计对于营造备受人们喜爱的环境是非常重要的。艺术是所有设计的基础。艺术家是视觉领域的研究者，并快速反映任何文化的现状和热点。艺术和设计密切相关，其可以表达设计思想，并作为知识分享和情感交流的手段。然而，正如人与人之间最强有力的联系纽带是情感，艺术和设计也在情感交流中最大限度地发挥作用。

最重要的是，作为设计师，我们创造美，而美却是难以定义或描述的，但同时它又是可以被大多数人认可的。美的定义因文化而异，但无论如何，在日常生活中，所有人都值得拥有美。

最后，本书中所收录的这些作品是多年来由玛莎·施瓦茨及合伙人设计事务所中众多才华横溢的设计师创作的。事务所内部的设计程序可以概括为"最好的设计思路是赢家"，而我的工作是帮助胜出的设计理念进一步发展并为其最终落成提供支持。得益于设计师们（我只是其中之一）的贡献，我们一直坚持运用最新的思路，我们的作品如此丰富多彩。设计特点之一是，没有"招牌风格"，所有人都不知道下一个设计会是什么样的。我们总是打造与场地条件和客户要求相适应的独一无二的景观，每一个设计都讲述一个特别的故事。

never reach global sustainability. There are simply too many people living in urban environments to neglect the human part within the sustainability equation. For optimum effectiveness, cities plans must be on an environmentally based master plan where low-carbon/renewable energy goal along with the creation of environments where human scale, needs and communities are created. Uses must be intermingled to lessen the pressure on transport. In other words, people must come before traffic planning.

Within the texture and scale of a city, we must go back to those environments that are nurturing to our human needs and behaviors. The neighborhood is a walkable domain where it is possible to shop, make important social connections, and walk or bicycle to work instead of being isolated in towers and dependent upon cars. People should be able to easily and safely cross roads and not be overwhelmed by the speed, noise and pollution of cars. Neighbourhoods are connected to other neighbourhoods and districts with tree-lined pedestrian-scaled sidewalks and bikelanes, which in turn, t into a larger environ- mental framework and infrastructure. Important for their quality of life, people need to shelter and delight from the stresses of city life within multiple green and open spaces. Lastly, people strive to nd beauty within their lives. All of these characteristics will help to create healthy cities where people will choose to live. A well-design, landscape-driven master plan is our biggest hope of creating this kind of sustainable environment.

In Landscapes for People, written by Garrett Eckbo in 1950, Eckbo focused on the relationship between creativity in the landscape and social interactivity, a topic that has been left out the recent discourse in landscape architecture. In our desire to expand beyond the site-scale and deal with landscapes at an urban scale, we seem to have lost sight of the human scale and the value we can create through the connections we make between each other through the emotional content expressed in art and design. Garret Eckbo and Lawrence Halprin, two of our founding fathers of the modern profession of landscape architecture, were humanists. They understood that we must keep our focus on people while living in balance with nature, and create places that will be loved, embraced and of cultural importance. This debate has evolved due to the expansion of the breadth of our profession. We should, theoretically, be capable of designing at all scales. However, while we must deal with urbanism at a larger scale and should learn to integrate information and form multi-layered strategies to solve complex urban issues. We must not forgo the importance of physical design at the human scale. At this scale human expression will be found and care about detail, all of which will broadcast quality to the user. People will see whether care has been taken to create a place. This, in turn, will be absorbed in one's esteem and identity. This all happens at a human scale, and people will know if the site has been designed with respect, humor and individuality. At the human scale a design can tell a story, create meaning, connection and create value to the people who come in contact with a space. Without a design that people value, we cannot achieve sustainability.

Quality design is essential to creating an environment that people will respond to. Art is the foundation for all design. Artists are the researchers of the visual realm and reflect what is immediate and topical in any culture. Design and art are in close conversation and are capable of expressing ideas and a means of communicating on an intellectual and emotional level. However, it is through our emotions that we most strongly connect to

前言 / Introduction 13

我们的设计实践集中在景观设计、城市环境及场所营造等方面。关于如何把可持续性设计理念应用于城市环境之中，得益于设计经验的逐渐丰富以及对设计理念的深刻理解，我们善于扩展、规划和设计大型场地，并将具有重要意义的生态系统融入其中，使景观具有可塑性和吸引力，使人感到愉悦。同时，我们也对临时性的艺术装置（比如，大型滨水项目）充满热情。在城市规划项目中，我们是很好的合作伙伴，为项目规划各种开放空间和公共空间系统。我们有能力领导大型团队并通过与其他领域专家的合作，收集如基础工程、社会和文化信息之类的重要资料，保证景观充分发挥生态功能，为人们打造可供使用、娱乐并乐意前往的场所。我们工作的目的是打动人心，激发人们去思考、感悟、猜想、享受，并把美和欢乐带入日常工作和生活。

玛莎·施瓦茨

each other. It is at the emotional level where art and design are at their most powerful.

Most importantly, as designers, we can create beauty, a quality that is arguably too difficult to define and therefore talk about, but is something that all people can recognize. Beauty is a quality that ranges from culture to culture but is, none the less, a quality that all people respond to and deserve to have in their lives.

Lastly, the work in this book has been created through the efforts of many talented people who have worked at Martha Schwartz Partners through the years. Our in-house process is "the best idea wins", and my job is to support and help to develop that idea through to its construction. Our designers (I am but one of them) are major contributors in ensuring we have continually fresh and new ideas to work with; we have such a richness of variety and expression in our work. One of the characteristics of our practice is that we do not have a house-style and it's anyone's guess what ideas will be generated next. We always produce a one-of-a-kind landscape that suits its particular site and client and tells a particular story.

Our practice remains focused on design, urban environment and place-making. With our expanding understanding of how sustainable practice is deployed in urban environments, we have also grown to be scalable, to be able to plan and design larger sites where we can insert meaningful ecological systems while still designing places that will attract and delight people. We are as enthusiastic about working on temporary art installation as much as large-scale waterfronts project. We work collaboratively on master plans, representing the open space or public realm systems. We can lead large teams, working with other experts so we can harvest fundamental engineering, social and cultural information critical to ensuring that the landscape will function ecologically while reaching the larger goals to create places that people will use, enjoy and to which they will return. The intent of our work is to touch people and invite them to think, feel, wonder, have fun and to bring beauty and pleasure in their lives.

Martha Schwartz

| 目录 | Contents | 14 |

序	Foreword	4
前言	Introduction	9
城市景观	Urban Landscapes	16

艺术作品 / Art Commissions

面包圈花园	Bagel Garden	20
轮胎糖果花园	Necco Garden	24
拼贴花园	Splice Garden	28
金县监狱广场	King County Jailhouse Plaza	32
迈阿密国际机场隔音墙	Miami International Airport Sound Wall	36
斯波莱托艺术节景观	Spoleto Festival	38
布劳沃德县市民体育场	Broward County Civic Arena	42
输电线景观	Power Lines	46
明日之城：Bo01住宅示范区	City of Tomorrow Bo01	52
花园装饰品	Garden Ornaments	54
铝酸盐景观	Aluminati	58
"城市与自然"主题花园	City and Nature Master Garden	64

住宅景观 / Residential

联排住宅	Nexus Housing	74
迪肯森住宅	Dickenson Residence	78
戴维斯住宅	Davis Residence	84
日本岐阜北方公寓	Gifu Kitagata Gardens	92
保罗林克庭院	Paul-Lincke-Höfe	100
纳提克努韦勒屋顶花园	Nouvelle at Natick	106

酒店与旅游胜地景观 / Hotels and Resorts

德拉诺酒店	Delano Hotel	112
迪斯尼乐园东入口广场	Disneyland East Esplanade	116

商业景观 / Commercial

里约购物中心	Rio Shopping Center	124
城堡购物中心	Citadel Shopping Center	130

医院景观 / Health Care

维也纳北方医院景观	Vienna North Hospital	136
卢森堡南部医院景观	Sudspidol Luxemburg	140

市政景观 / Civic

HUD 广场	HUD Plaza	146
明尼阿波利斯联邦法院广场	Minneapolis Courthouse Plaza	152
雅各布·贾维茨广场	Jacob Javits Plaza	156
林肯路购物中心	Lincoln Road Mall	160
交易广场	Exchange Square	164
佛里斯顿村庄绿地	Fryston Village Green	170
梅萨艺术中心	Mesa Arts Center	174
大运河广场	Grand Canal Square	182
贝鲁特滨水公园设计竞赛	Beirut Waterfront Park Competition	190
索沃广场	Sowwah Square	194
海军码头公园设计竞赛	Navy Pier Competition	200
共和广场	Place de la Republique	206
北钓鱼台开发项目	North Diaoyutai Development	212
莫斯科儿童休闲街区	Moscow Children's Route	216

| 目录 | Contents | 15 |

公司办公景观	Corporate

技术创新中心	Center for Innovative Technology	222
贝顿迪肯森公司总部	Becton Dickenson Headquarters	226
瑞士再保险总部大楼	Swiss Re Headquarters	230
巴克莱银行总部大楼	Barclays Bank Headquarters	236
万科中心	Vanke Center	242
北七家科技商务区	Beiqijia Technology Business District	248

公园	Parks

滨海线性公园	Marina Linear Park	258
约克维尔公园	Village of Yorkville Park	264
蒙特拉中央公园	Monte Laa Central Park	272
圣玛丽教堂公园	Saint Mary's Churchyard Park	282
凤鸣山公园	Fengming Mountain Park	286

再生景观	Reclamation

| 温斯洛农场保护景观 | Winslow Farm Conservancy | 296 |
| 麦克劳德尾矿 | McLeod Tailings | 304 |

总体规划	Masterplans

多哈滨海大道设计竞赛	Doha Corniche Competition	310
露露岛	Lulu Island	316
龙山国际商务区	Yongsan International Business District	320
阿布扎比滨海沙滩公园	Abu Dhabi Corniche Beach	324
普鲁伊特市规划	Pluit City	328

附录	Appendix

奖项	Awards	336
设计团队	Team	337
设计师小传	Biographies	338
参与人员名单	Credits	342

城市景观　　　　　　　　　　　　　　　　　　　　　Urban Landscapes

在当今的星球上，有一半以上的人口生活在城市之中。大规模的全球城市化推动了新城市和旧城市迅速发展的大潮。然而，对城市的运行、适宜居住的能力以及未来一代的可持续发展起决定性作用并为城市发展提供助力的，正是城市景观设计。人们逐渐认识到，在全球市场竞争的大环境下，城市景观已经迅速发展成一个影响城市健康发展、环境友好、经济价值创造并吸引投资以及促进文化发展等的关键因素。

在传统意义上，大型城市景观可以满足城市环境营造和人类健康的需求，以及人们对社会和集体空间的需求。良好的社会和集体空间有助于营造积极的城市环境，使社区更令人向往，更具独特性。玛莎·施瓦茨及合伙人设计事务所的作品体现了对这种需求的深切关注。施瓦茨认为，现代城市景观必须超越传统象征学的局限性，把城市生活中的各个方面，特别是那些具有利用价值并代表城市特征的空间，纳入其中。在城市中，完整的景观概念需要进一步扩展，它不仅包括公园或者屋顶绿化，还包括街道、步行道、设施廊道、停车场以及建筑外部的一切——实际上，这就是人们花费大部分时间停留和活动的地方。

然而，正是这些被低估了的城市空间构成了公共领域，并作为当今的城市生活平台。这些存在于建筑之间的空隙结构，为人们提供了各种可供非正式的聚会、再创造以及其他活动的场所和空间。玛莎·施瓦茨及合伙人事务所通过多年来的设计实践，把这些空间改造成颇具人性化且充满想象力的场所，使公共空间和私人领地充满生机和活力，为在城市中工作和生活的人们带来惊奇、美丽、自然和欢乐。事务所实践的基本信条是：让人们积极参与并充分满足人们的基本需求，比如，相互之间的

More than half of the planet's population is now living in cities. This vast global trend towards urbanization that is underway encompasses a surge of new and old metropolises that are in the process of developing. Decisive to the city's performance, liveability and sustainability for future generations and underwriting this development is the urban landscape. It is rapidly becoming understood as a crucial factor in the ability of a city to provide the kind of healthy domain that will attract people, be environmentally friendly, create economic value, attract capital and contribute to a city's cultural advantages in a globally competitive market.

Traditionally, great urban landscapes have helped to fulfil the needs for environmental and human health, for social and collectivized urban space that generates a positive quality of life in cities that result in making these communities desirable and unique. The work of MSP demonstrates a deep commitment to this need yet raises the ante to propose that the contemporary urban landscape must go beyond the traditional typologies and embrace all aspects, sectors and territories of urban life, especially the utilitarian spaces that characterize the city. The whole idea of landscape in the city needs to be expanded beyond the normal idea of parks or green roofs. We must consider that most of our urban environments consist of streets, sidewalks, utility corridors, parking lots and everything outside the buildings. This is, in fact, where people spend most of their time.

Yet, it is these undervalued urban spaces that constitute the public realm and serve as the platform for urban life today. Within the interstitial fabric of in-between spaces of buildings exists a panoply of diverse places and spaces that have the potential for citizens to informally meet, recreate and connect with one another. MSP, through their design practice, urges these settings to become humane and imaginatively detailed so that, in the process, we positively activate public and private territo-

城市景观

联系、个性化表现和娱乐休闲——这些是构成景观重要性最深层次的原因。人们应该拥有各种不同的体验，它们并区别于邻里社区、不同的场所和不同的场地，特别是在全球化日益深入和现有世界同质化日益加剧的背景下。

另一个让景观设计在城市营造中占据支配地位的原因是当今城市发展对于环境问题和可持续发展的关注。许多大型景观项目能够降低极端高温，促进水的循环，对降水进行收集和再循环使用，减少能源消耗，减少碳排放，吸引野生动物，改善居民的健康状况和城市生活的不利因素。许多新技术被用于应对环境问题；生态学的相关理念也被景观设计专业所接纳。然而，对于一个景观设计师来说，无论其采用的技术多么先进，或从生态学的角度多么合理，如果人们感觉不到与城市景观的密切联系，那么它都不算是一个成功的设计。出于这种原因，玛莎·施瓦茨及合伙人设计事务所的作品在非常看重设计的艺术表现的同时，也同样关注可持续发展和生态因素。设计可以营造某种场所感和归属感，让人们通过使用和亲身体验，与场地建立情感连接。此外，设计还可以强化人类与自然环境之间的关联，在二者之间营造一种巧妙的平衡。

为了达到这种平衡，玛莎·施瓦茨及合伙人设计事务所进行了一系列探索，并成功打造了许多高效且实用的公共空间和私人空间，它们构成了新型公共景观并组成了城市文化的展示舞台。事务所把城市文化在景观设计中的体现理解为景观设计在空间设计层面的作用：个人和社会塑造了文化愿望——他们想要看到自己，同时也让自己被世界看到。事务所的作品横跨景观、建筑、装饰性雕塑、园艺、工程、生态学、科学、技术以及视觉艺术等多个学科，并把它们按照不同的比例巧妙搭配，使每一个项目都成为符合城市愿望的"场地"。在设计过程中，人们经常向事务所中的景观设计师询问究竟什么样的形象才能使一个项目、一个社区甚至一个城市实体彰显其独一无二的社会、文化和民族特征，并最终被公众接受。结果是，事务所承接的景观、装置和大型城市项目一直在为客户营造全新的空间和环境。

现在，城市健康发展的重要标志之一就是该城市是否对寻求高质量生活的人具有吸引力。在过去，人们对于城市美化的认识是他们对绿色空间和林荫道路的可接近程度——这曾被用来吸引知识分子前来居住和工作。然而，鼓励人们生活在一起，积极合作，实现阶级整合，在能源、食品和交通运输方面减少对自然资源的消耗，对于实现全人类的可持续发展更是至关重要的——这就是玛莎·施瓦茨及合伙人设计事务所所坚信的"城市化进程未来的发展方向"。

马库斯·詹斯奇和伊迪丝·卡茨

Urban Landscapes

ries with design qualities that bring delight, beauty, nature and playfulness to the city. Fundamental to the practice is the belief that the engagement of people and satisfaction of basic human needs for connection, identity and enjoyment underlie one of the deepest reasons why landscape is becoming so important. As humans we need to experience various identities and differentiation between neighborhoods, locales and individual sites, especially in the face of increasing globalization and the tendency towards homogenization of our built world.

The other reason that landscape is in ascendancy is due to environmental concerns and issues of sustainability in the city. Great landscape design can perform functions that mitigate extreme heat, moderate the hydrologic cycles, harvest and recycle storm water, reduce energy usage, lower carbon emissions, attract wildlife, benefit the health of residents and help ameliorate the harsher aspects of urban life. Many new technologies have emerged to address environmental issues and ecology has been embraced by the profession. Yet, no matter how cutting-edge the technologies employed or how ecologically correct a project may be, if people don't feel connected to that urban landscape it will not be successful. For this reason, MSP places a great emphasis upon the artistic expression of design that must be appreciated as a vital factor while also considering sustainability and ecological factors. Design can create a sense of place and engender, a sense of belonging and individuality that encourages an emotional connection to a place by the people who use and then experience it. Design can also improve the platform upon which human and natural environments can be brought into an artful balance.

MSP's explorations in achieving this balance has led to the design of highly effective and well used public and private spaces. These public and private spaces constitute a new public realm landscape that is the stage for urban culture. MSP understands the expression of urban culture in the landscape as the space of appearance: where individuals and the society convey in built form the cultural aspirations as they wish to see themselves and be seen by the world. MSP operates transversally across landscape, architecture, installation sculpture, horticulture, engineering, ecology, science, technology and visual art deploying these modalities in varying ratios, as each project requires, to create a "place" that will meet the cultural aspirations of the city. During the design process, MSP is often asked to decipher what the image should be for an individual project, a community or even an urban entity that responds to its unique social, cultural and ethnic characteristics so that it will be embraced by the public. Consequently, the team's landscapes, installations and large urban projects continue to shape the environments they are part of and the lives of the people who use them.

This is now of utmost importance to the health of any city as it positions itself to be an attractor to people seeking a certain quality of life. Beautification as we have known it in the past, in the sense of accessibility to green spaces and tree-lined streets, is being used to entice knowledge-based workers to come to live and work in the city. Encouraging people to live together, collectively, mixing social classes, using less resources spent on energy, food and transport is vital to achieving a sustainable planet and MSP believes in the urbanization process that is underway for this reason.

Markus Jatsch and Edith Katz

艺术作品   **Art Commissions**

| 艺术作品 | Art Commissions | 20 |

面包圈花园

美国马萨诸塞州波士顿，1979

作为一名训练有素的艺术家，施瓦茨曾有过在剑桥的一家景观建筑公司饱受挫折的学徒经历。寄希望于快速且低成本的手工作品，她对自家坐落于后湾的乔治风格联排别墅前院中的一个花园进行了设置。结合花园原有的两个同心方形绿篱规则式花园设计，施瓦茨将法国文艺复兴风格的花园设计应用其中，营造了一个可供舞会和活动庆祝的舞台布景。

除了作为惊喜聚会的场所，这个前院同时包含景观艺术家对于其钟爱的食物的赞美："面包圈是谦逊的、有家庭氛围的、具有民族特色的。"她解释道，"另外，我用并不昂贵的价格就可以得到很多。"在一内一外两个由40.6厘米高黄杨木组成的方框之间，施瓦茨设计了一个76.2厘米宽且主要供水族箱使用的紫色碎石铺就的宽带区域，并用8打面包圈在上面制作了网格点阵。每个面包圈都浸泡过海洋石清漆，以防风雨侵蚀。在内侧的树篱方框中，她种植了30株紫色藿香，与紫色碎石和原有的日本枫树形成呼应。

该作品被刊登在《景观建筑杂志》上，并成为1980年1月刊的封面故事。之后，该设计引发了业内对"景观设计作为文化表达形式的能力"的激烈讨论。作为第一个"概念性景观"，这个花园成为业内的一个历史性起点。

Bagel Garden

Boston, MA, USA (1979)

Trained as an artist, Schwartz grew frustrated working as an apprentice in a Cambridge landscape architectural office. Longing for a speedy, inexpensive installation she could accomplish with her own hands, she tackled the front garden of her own Georgian row house in Back Bay. Incorporating two concentric square hedges of an existing formal garden, Schwartz based her scheme on French renaissance gardens, which were designed as stage sets for dances and celebrations.

Conceived as the stage set for a surprise party, the front yard also became an ode to the landscape artist's favorite food: "Bagels are humble, homey, and ethnic," she explains. "Besides, I could get many of them inexpensively." Between the outer and inner squares of the 16-inch high boxwood hedges, Schwartz arranged a 30-inch wide strip of purple aquarium gravel dominated by a grid of eight dozen bagels. Each bagel was dipped in marine spar for weatherproofing. Inside the inner square of hedge, she planted 30 purple Ageratum to match the gravel and complement an existing Japanese maple.

The garden was then published by *Landscape Architecture* Magazine and appeared as the front cover story in January 1980. What followed was a heated debate within the profession about the ability of the landscape to be a form of cultural expression. It became a historical point within the profession as the first "Conceptual Landscape".

艺术作品 Art Commissions 21

艺术作品　　　　　　　　　　　　　　　　　　　　Art Commissions　　　　　　　　　　　　　　　　　　　22

艺术作品　　　　　　　　　　　Art Commissions

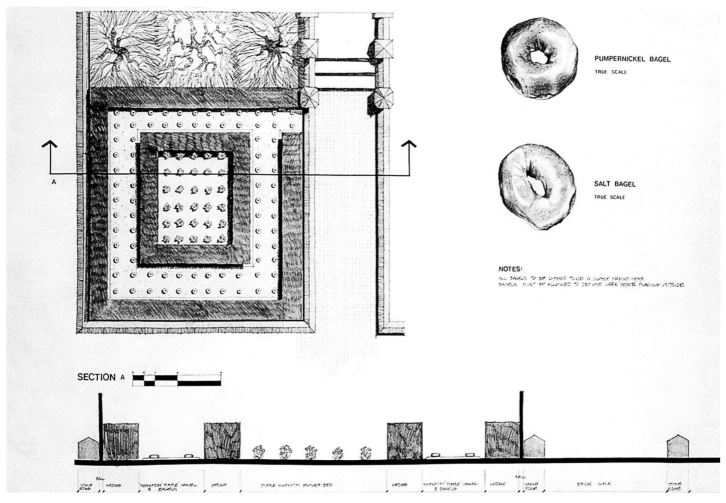

| 艺术作品 | Art Commissions | 24 |

轮胎糖果花园

美国马萨诸塞州剑桥，1980

轮胎糖果花园，秉承伟大的古典法式花园的精神组织了平面图案，并通过对连续排列和平行线的应用，扩展了距离。几何机构设计，不仅彰显出大中庭作为麻省理工学院的象征的宏伟和庄严，还使人联想起孩提时代单纯的快乐，并畅享春季伊始的美景。花园在配合庭院规模和秩序的同时，映衬着从查尔斯河岸边可以遥见的波士顿后湾网格状大型城市背景。

Necco Garden

Cambridge, MA, USA (1980)

The Necco Garden was done in the spirit of the great classic French gardens and organized by patterning the horizontal planes, by using objects in a serial fashion, and through the use of parallel lines to exaggerate distances. The design, through geometry, reflected the grandeur and formality of the Great Court (which is symbolic of MIT), while evoking simple pleasures of childhood and the joy that accompanies the first signs of spring. The garden responded to the great scale and order of the courtyard while reflected the larger urban context of the grid of Back Bay Boston which is visible across the Charles River.

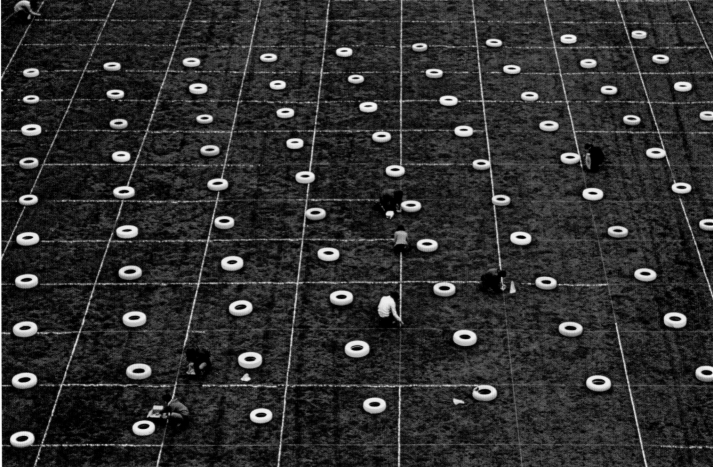

艺术作品　　Art Commissions　　27

| 艺术作品 | Art Commissions | 28 |

拼贴花园

美国马萨诸塞州剑桥，1986

拼贴花园位于马萨诸塞州剑桥，长10.7米，宽7.6米，是由怀特黑德研究所——一个微生物学研究中心的所长大卫·巴尔的摩收集的"敢于冒险的艺术收藏品"的一部分。该项目位于一个由波士顿建筑公司古蒂克兰西设计事务所设计的九层高死气沉沉的办公楼楼顶。单调乏味的瓷砖屋顶表面和高高的围墙，构成了一个黑暗且荒凉的空间。人们可以从楼上的一个教室和一个职工休息室鸟瞰，并从职工休息室进入屋顶庭院。这使该庭院成为潜在的午餐场所。

屋顶庭院的空间并不宽敞，同时屋顶地面采用混凝土板结构，无法负担额外的重量。此外，屋顶上没有水源，没有管护人员，项目预算很少——这基本排除了种植植物的可能性。然而，运用足够多的象征性元素使它看起来像一个花园，并给人一种"种满植物的花园"的感觉，这是完全可能的。在世界各地，有很多创造"虚幻花园"的实例。例如，在日式花园中，象征性景观常常被用来展示更大的景观。怀特黑德研究所的计划是：通过抽象、象征和引用的方式创造一个花园。同时，施瓦茨希望这个花园的设计语汇与怀特里德研究所所从事的工作相关联。最终，这个花园讲述了一个警示性的故事，提醒人们基因剪切的内在危险性：它可能创造出"怪物"。

这个花园就像一个"怪物"——一个结合了双重文化的"连体婴儿"。一边是基于法国文艺复兴风格的花园；另一边是日式禅宗花园。然而，这两个花园的构成要素已经发生变形。日式禅宗花园中常见的岩石变成了法式花园中的造型灌木。其他植物，比如棕榈和松树，也以一种奇特或不寻常的方式组合在一起。有些植物探出墙面，垂直于墙体，还有一些植物似乎在墙顶摇摇欲坠。

花园中的所有植物都是塑料的。修剪成型并同时可用作座椅的绿篱，由轧钢覆以阿斯特罗特夫尼龙草皮制成。绿色，作为体现这是一个花园的最强烈暗示，由绿色砾石和喷漆表现出来。这样做的目的是在这座大楼里工作的科学家创造一个无法解决的视觉难题。这个花园是对"化学改变生活"的歌颂。

Splice Garden

Cambridge, MA, USA (1986)

This 25-foot by 35-foot rooftop garden in Cambridge, MA, is part of an adventuresome art collection assembled by Director David Baltimore for the Whitehead Institute, a microbiology research center. The site was a lifeless rooftop courtyard atop a nine-story office building designed by Boston architects Goody Clancy Associates. Its dreary, tiled roof surface and high surrounding walls conspired to create a dark, inhospitable space, overlooked by both a classroom and a faculty lounge. The lounge offered access to the courtyard, making it a potential place to eat lunch.

Along with its spatial woes, the floor of the courtyard was constructed with a concrete decking system that could not hold additional weight. There was also no source of water for the rooftop, no maintenance staff, and a low budget, precluding the possibility of introducing living plants. However, it was entirely possible to convey a sense of a planted garden by providing enough signals for the site to read as a garden. There are many examples of other cultures that create garden abstractions. For example, in Japanese gardens, symbolic landscapes often imply a larger landscape. This was the strategy at Whitehead to create a garden through abstraction, symbolism, and reference. Schwartz wanted the narrative of the garden to relate to the work carried out by the Institute. The garden became a cautionary tale about the dangers inherent in gene splicing: the possibility of creating a monster.

This garden is a monster—the joining together like Siamese twins of gardens from different cultures. One side is based on a French Renaissance garden; the other on a Japanese Zen garden. The elements that compose these gardens have been distorted. The rocks typically found in a Zen garden are composed of topiary pompons from the French garden. Other plants, such as palms and conifers, are in strange and unfamiliar associations. Some plants project off the vertical surface of the wall; others teeter precariously on the wall's top edge.

All the plants in the garden are plastic. The clipped hedges, which can be doubled as seating, are rolled steel covered in Astroturf. The green colors, which are the strongest cues that this is a garden, are composed of colored gravel and paint. The intent was to create, for the scientists who occupy this building, a visual puzzle that could not be solved. The garden is an ode to "better living through chemistry".

艺术作品 Art Commissions 29

艺术作品　　　　　　　　　　　　　　　　　　　　　　Art Commissions　　　　　　　　　　　　　　　　　　　　　30

艺术作品　　　　　　　　　　　　　　　　　　　　　　　　**Art Commissions**　　　　　　　　　　　　　　　　　　　　　　　　31

| 艺术作品 | Art Commissions | 32 |

金县监狱广场

美国华盛顿州西雅图，1987

西雅图新监狱，没有专门设计犯人与律师、带孩子的家庭以及其他来访者会面的大厅或休息室。取而代之的，是由混凝土和瓷砖构成的广场花园，提供了一个充满活力并且维护成本低的聚会场所。景观雕塑的形式多样，这里可设置儿童娱乐设施、回声绿篱、花坛以及喷泉。广场地面采用破碎的瓷砖与露骨料条带交替的铺装。广场后面建筑墙体上的陶瓷壁画暗示着开放空间；简单的拱形结构作为花园的大门，表明有出口。尽管从图形上看去有"逃跑"之意，但实际上，瓷砖墙体仍然清楚地表明这是监狱。

该设计方案力求把焦点集中在地平面上，而不是这座令人感到压抑的建筑本身。等人身高的景观元素及其色彩和为原本萧索的空间带来了几分舒适感。众多"造型景观雕塑"都可用作座椅。破碎的瓷砖昭示这个花园正处在分崩离析的边缘。彩色色调和设计主旨传达了一个更深层次的信息：对犯人混乱、危险和脆弱的生活的关注。

King County Jailhouse Plaza

Seattle, WA, USA (1987)

Seattle's new jail, designed without a lobby or a foyer, offered nowhere for attorneys, families with children, and other visitors to meet. A plaza garden of concrete and ceramic tile offers a lively, low maintenance meeting place. Sculptural forms, serving as children's play structures, echo hedges, topiary, parterres, and a fountain. Its surface is paved in broken tile alternating with stripes of exposed aggregate. A ceramic mural on the wall of the building behind the plaza suggests open space; the simple arch represents a garden gate, implying an exit. Although the image suggests the idea of escape, the tiled wall in fact imprisons.

The scheme is intended to focus attention on the ground plane and away from the overwhelming bulk of the building. The color and human scale of the objects bring comfort to a harsh, cold space; the "topiary" forms provide places on which to sit. The tile fragments suggest that the garden is at its penultimate moment before disintegration. The colorful palette and subject matter belie a deeper message: a recognition of the chaos, danger, and fragility of prisoners' lives.

艺术作品　　　　　　　　　　　　　　　　　　　　　　　　**Art Commissions**

艺术作品 | Art Commissions

艺术作品　　Art Commissions

| 艺术作品 | Art Commissions |

迈阿密国际机场隔音墙

美国佛罗里达州迈阿密，1996

沿着迈阿密国际机场北边界和第 36 大街，有一道长 1.6 千米的隔音墙，彩色玻璃和射在其上的明媚的阳光使隔音墙显得生机勃勃。这道隔音墙把机场与两个邻里社区分隔。隔音墙由预制混凝土板制成，高 7.6 ~ 10.7 米。设计任务是改造这道不受周边社区居民喜欢但为机场所必需的隔音墙，使周边社区居民也喜欢它。

墙体面朝北方，面向第 36 大街的立面总是处于阴影之中。因此，喷漆、贴布绣或者安装浮雕都是不可取的。最终，设计方案决定用阳光本身为这道墙体增添活力；在墙上凿出洞，并镶嵌彩色玻璃，阳光折射的彩色光圈使隔音墙充满生机；六种造型不同的面板形成了随机的图案；对墙体顶部进行重新塑造，对墙体底部景观进行平整处理。这道隔音墙沿着 1.6 千米长的道路，起伏变幻。在机场出口处，这些孔洞组合成网格形式。这种变化使这些孔洞看起来具有脉动感，象征着运动。

Miami International Airport Sound Wall

Miami, FL, USA (1996)

Colored glass and sunlight enliven a mile-long sound attenuation wall along 36th Street which lies along the northern boundary of the Miami International Airport. Separating the airport from two adjacent neighborhoods, the wall is constructed of precast concrete panels and ranges in height from 20 to 35 feet. The design task was to make the concrete barrier wall, which the neighborhoods didn't want but the airport deemed necessary, into something the neighbors could enjoy.

Because the wall faces north, the facade along 36th Street is always cast in shadow. For this reason, solutions such as painting or appliquéing the wall with a bas-relief were rejected. Instead, in the final design, the sun itself is used to energize the wall. Holes inset with colored glass are punched between the steel reinforcing bars of the wall and allow the sun to create circles of colored light. Six different panels were designed to create the appearance of a random pattern. By reshaping the top of the wall and regarding the landscape at its base, the sound wall undulates down the mile-long stretch of roadway. At the entrance to the airport, the holes organize into a grid. This change in pattern makes the holes appear to pulsate and implies motion.

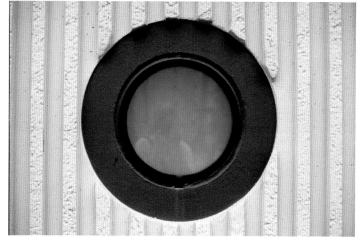

| 艺术作品 | Art Commissions | 38 |

斯波莱托艺术节景观

美国南卡罗来纳州查尔斯顿，1997

以下内容截取自文章《致编辑的一封信》（印刷：南卡罗来纳州查尔斯顿邮政快递局，1997年6月19日）。

弗兰克 W. 克莱门特的短信（主题：种植"艺术"，6月1日）的结尾提出了这样一个问题："这可以被称为'艺术'吗？"我感到有必要给以回答。

我在福来路上开车向北行驶，经过麦克劳德种植园，第一次见到玛莎·施瓦茨的作品。我曾经听到有人开玩笑地称它为"玛莎之洗"，我当时赞同这个称呼。然而，当我步行穿过这个种植园的时候，我的想法改变了。

我的第一感觉是，该作品真正的标题应该是"野外之作"。它让人立即联想起曾经围绕麦克劳德的大片棉田。但是，我无法想象，环绕在我身边随风摆动的大块白色布单与棉田有什么关系。施瓦茨真的想要体现"洗衣服"的主题吗？把布单晾在一片田野之中，让它晒干？她为何这样设计？

我站在两行晾晒的布单之间，看到一间木质小屋，找到了以上问题的答案。小屋呈风化的灰白色，由两个房间组成，每个房间中居住着一个大家庭。奇怪的是，晒衣绳之间的小路被漆成白色。小屋之外有类似的白色草坪，白色布单漂浮其间。

在非洲艺术中，白色象征着"精神世界"。当我意识到这一点时，我突然发现，自己处于施瓦茨所编织的"精神世界"之中。透过余音，听到黑人妇女微弱的声音，日复一日、年复一年地在种植园里，把洗过的东西挂上拿下，一边聊天，一边歌唱。

透过余光，我看到正在捉迷藏的黑人小姑娘模糊的身影，就在晒衣绳之间，将来长大接替母亲的工作。随着观察的深入，这些布单又有了些许新意。有些变成以前大型轮船的船帆，把非洲人运到查尔斯顿当奴隶；有些变成鬼精灵，附着在这片饱受苦难的土地上，无法离开。

在我看来，施瓦茨的作品感人至深，并对那些在麦克劳德的小木屋中辛勤劳作的黑人妇女表达了真诚的敬意。这些作品通过一系列生动形象的视觉影像，向人们展示了一个真实且美丽的"精神世界"。

威尔默 H. 威尔士

Spoleto Festival

Charleston, SC, USA (1997)

The following is an article reprinted from the "Letters to the Editor" in the Charleston, South Carolina Post and Courier, June 19, 1997.

Frank W. Clement's brief letter (Plantation 'Art', June 1) ended with a question: "How can we call this 'art'?" I felt compelled to offer an answer.

I was driving north on Folley Road past the McLeod Plantation when I first saw Martha Schwartz's piece. I had heard it jokingly called "Martha's Wash", and I agreed with the title. However, that changed later as I walked through it.

My first clue as to its meaning, at least for me, was its real title "Field Work". That immediately suggested the cotton fields which at one time surrounded McLeod. But I could not think of any relationship between cotton fields and the large sheets of white fabric that flapped in the wind around me. Had she really meant them to look like wash, which is certainly work, but of another kind, hung out in another kind of field to dry? If so, why?

I found the answer to those two questions when I stood at the bottom of an alley of wash looking up at one of the wooden slave cabins. Weathered whitish grey, it consisted of two rooms, in each of which an entire family had lived. Oddly the grass in the alley was painted white. All of the cabins had similar alleys of white grass between white, flapping sheets.

In African art, white is the color reserved for the spirit world. And as I recalled this, I suddenly found myself in the midst of a spirit world conjured up by Martha Schwartz. Out of the corner of my ear, I could hear the faint voices of black women talking and singing as they hung up and took down the plantation's wash, day after day, year after year.

Out of the corner of my eye, I could see the dim shapes of little black girls playing hide-and-seek between the lines and growing up to take their mother's places. As I watched, even the sheets took on a new meaning. Some became sails of long-gone vessels that had brought the Africans to Charleston as slaves. Others were the ghosts of those who still clung to the site of their suffering, unable to let go.

To me Martha Schwartz's piece is a moving tribute to the persistent courage of generations of black slave women who humbly worked, gave birth and died in the cabins of McLeod and elsewhere in our beloved South. Like all good art, it points no fingers; it conveys its subtle messages with visual symbols to those of us who try to see.

Wilmer H. Welsh

| 艺术作品 | Art Commissions | 42 |

布劳沃德县市民体育场

美国佛罗里达州福特劳德代尔，1998

从前这里是佛罗里达大沼泽地的一部分，现在是汽车租赁中心、佛罗里达美洲豹冰球队的所在地。布劳沃德县公共艺术与设计规划委员会计划把中心入口广场打造成一个综合性艺术作品展示区。概念设计的重点是大尺度雕塑般的树冠结构，它由钢框架和高科技材料制成。这16个人工树冠让人联想起从前的湿地树木，形状和色彩均参照皇家棕榈树。两排带有"机械棕榈"风格的棕榈树把人们引向中心入口。树冠采用内部照明，夜晚灯光闪烁，照亮通往中心的步行道。条形铺装格局与建筑几何形态相互映衬，对广场上的树冠结构起到强化作用。

Broward County Civic Arena

Fort Lauderdale, FL, USA (1998)

On a site formerly part of the Florida Everglades rises the Broward County Civic Arena, home of the Florida Panthers Ice Hockey Team. The Broward County Public Art and Design Program commissioned this site—integrated art work for the Arena's entry plaza. The concept developed for the site includes the design of large scale sculptural canopies made of steel framing and high-tech fabric. These sixteen canopies recall the displaced Everglade trees, and, taking their cues for form and color from Royal Palms, the two rows of stylistic "mechanical palms" lead the way to the Arena's entry. Lit from within, the canopies glow at night and light the path of pedestrians on their way to Arena events. A striped paving pattern, echoing the building's geometry, reinforces the arrangement of the canopies in the plaza.

艺术作品   **Art Commissions**   43

| 艺术作品 | Art Commissions | 46 |

输电线景观

德国盖尔森基兴，1998

"梅希腾贝格"是艾姆歇采矿区唯一一座自然山脉。在这一地区，电力的汇集涉及几个因素：在神话方面，有关沃坦神（北欧神话中的众神之父）的神秘历史；在政治方面，俾斯麦纪念碑；在经济和环境方面，能源工业的动力线。山坡上与电线平行设置的线性几何结构玉米田有助于强化地形特征。

"红色走廊"由干草堆集的两道墙体围合而成，是俾斯麦纪念碑的一个组成部分，一直延伸至电力线的中轴线。走廊被漆成红色，代表电力的颜色和血的颜色。干草堆之间的通道非常狭窄，当两个人相遇时，需要考虑一下"谁先通过"，这条狭窄的通道象征着电力工作充满令人无法预知的困难。

"黑色空间"位于俾斯麦纪念碑与电力中轴线交叉处。这是一个圆形空间，由干草堆成，外面由黑色塑料袋包裹。地面采用煤炭铺装。这个"黑色空间"也被称为"黑色心脏"或作为景观中心。在这个空间中，人们可以沉思冥想，想象一下世人为电力所付出的巨大代价，这种代价既包括政治上的，也包括环境上的。

Power Lines

Gelsenkirchen, Germany (1998)

"Mechtenberg" is the only natural hill in the coal-mining area of the Emscher region. On this site, several elements that deal with issues of power converge: on the imaginative level, the mythical history around the god Wotan; on a political level, the monument for the chancellor Bismarck; on an economic and environmental level, the power lines of the energy industry. To heighten the topography, a geometrical structure of linear cornfields is superimposed upon the hill. The linear structure of the corn is generated from and runs parallel to the electric power lines.

The "Red Corridor", defined by two walls made from stacked hay bales, marks the axis from the Bismarck monument to the power lines. The corridor is colored red, a "power" color and the color of blood. The pathway between the hay bales is very narrow, forcing visitors to think consciously about who can pass when two people meet. In a very direct and immediate way, it presents the difficulty of dealing with power.

The "Black Room" is located at the intersection of the Bismark and power line axes. It is a circular room contained by stacked hay bales wrapped in black plastic. The floor is made of coal. This room is the "Black Heart" or center of the installation. Within this room, one might contemplate the high price we pay for power, both politically and environmentally.

Art Commissions

| 艺术作品 | Art Commissions | 58 |

铝酸盐景观

冰岛雷克雅未克，2008

作为"雷克雅未克实验马拉松"各项活动的组成部分，这座大型户外艺术装置于 2008 年 5 月在雷克雅未克艺术博物馆揭幕。

这件艺术品好像一个巨大的黑盒子，宽 14 米，长 14 米，高 5 米。标题为"我痛恨自然 / 铝酸盐"。设计灵感源自社会上流行的一种错误观点：在当今世界，有无穷无尽的自然资源可供人们开发利用。

这个巨大的盒子位于博物馆的前院。游客进入其中，沿着走廊前行，会看到一系列框景，那是一些令人眩晕且皱巴巴的由工业铝材构成的空间。

内部空间似乎在形状和体量上难以描述，没有实际的参照物可供对比。漆黑的内部走廊与面朝天空的核心区令人眩晕的光线形成了鲜明的对比，但实际上，游客几乎看不到。铝映衬出冰岛白天的天气变化。

这件艺术品所反映的主题就是铝本身。在盒子内黑暗的环境中，铝让人感到昏昏欲睡甚至处于危险之中，但同时它又极具吸引力。铝，特别是它的熔炼过程，正是目前冰岛颇具争议的问题。铝本身是漂亮诱人的，但是，国家经济如果严重依赖于铝熔炼，那么将付出难以估量的代价。

这件艺术品生动形象地表明，在当今世界，人们对自然资源的属性和重要性日益关注，并已采取一系列措施，旨在以绿色、环保的方式保护并利用自然资源。人们已意识到，持续恶化的生态景观必将对人类社会的可持续发展产生极其严重的负面影响。

然而，在现实中，人们的思想与行为相分离，加之人们脑海中早已根深蒂固的"破坏自然"的错误观念，"共建美丽的自然家园，实现人类社会的可持续发展"仍将任重而道远。

Aluminati

Reykjavik, Iceland (2008)

This large outdoor art installation was revealed at the Reykjavik Art Museum in May 2008 as part of a wider event entitled the Experiment Marathon Reykjavik.

The piece, a huge black box of 14 m width × 14 m length × 5 m height is entitled "I Hate Nature / 'Aluminati'" and is inspired by society's "delusional view" that there are limitless natural resources to exploit in the modern world.

Visitors walk into this box, located in the front courtyard of the museum, and from the corridor inside, are presented with a series of framed views onto a blinding space of crinkled, industrial aluminium.

The interior space is seemingly indescribable in shape and size and has no real reference to scale. The blackened interior corridor is in sharp contrast to the dazzling light within the core —which is opened to the sky, but not really visible to the viewer. The aluminium reflects the changing quality of an Icelandic day.

The subject of the installation is the aluminium itself. From its position in the black environs of the box, it is both mesmerising and repulsive, attractive and dangerous—the aluminium, specifically the smelting operation to which it refers, is currently a subject of controversy in Iceland. Its beauty is seductive, belying the price the country will pay for its economic dependence on the smelting process.

According to the theory that inspired and accompanies the piece, it assesses the modern view of "nature" as being disconnected from the reality of contemporary landscape. A landscape that is overwhelmed with visual degradation has a consequently negative impact on sustainability.

The disconnection between what we say and what we do about nature, along with our misconception of our place in nature, disallows us from developing a proactive, form-giving attitude towards the built environment.

艺术作品 | **Art Commissions** | 59

艺术作品　　　　　　　　Art Commissions　　　　　　　　60

艺术作品   **Art Commissions**   61

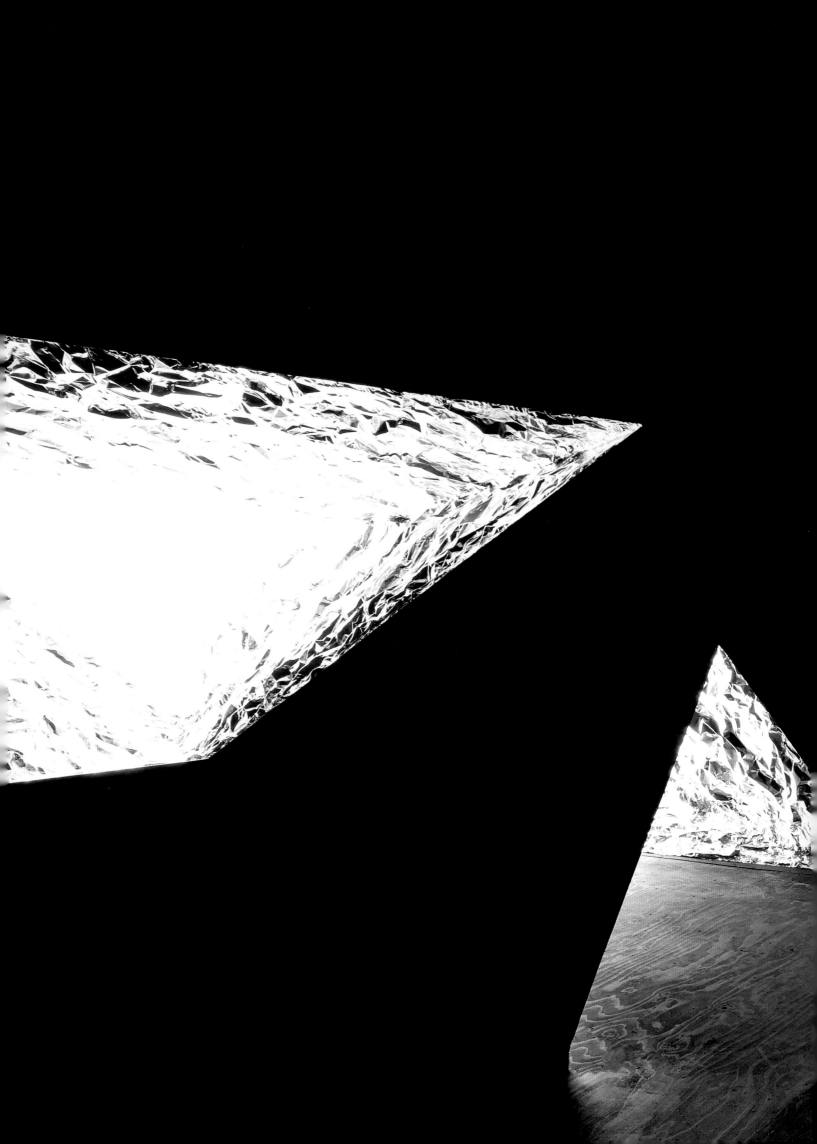

| 艺术作品 | Art Commissions | 64 |

"城市与自然"主题花园

中国西安，2011

2011年中国西安世界园艺博览会组委会邀请了9家国际景观设计公司，旨在围绕"城市与自然和谐共存"这一主题，设计一个小型花园。玛莎·施瓦茨及合伙人设计事务所是其中一家。根据组委会的指导思想，设计方案应该充分考虑当地建筑材料的实际情况和施工方法，这个花园应该是舒适、美丽且易于接近的。

这个花园的主题是"城市与自然"，主要由四个元素组成：传统的青砖墙和地面铺装，垂柳，单面镜和铜铃。美学设计灵感源自中国传统建筑及其与自然的密切联系。

长期以来，青砖一直是中国传统建筑的主要材料。青砖与木材已被使用了多个世纪。在城市空间营造和私有领地保护方面，这种墙体是最常见的构建元素，在宫殿中也经常被采用，并且是权力的象征。对于世世代代居住在庭院建筑中的许多人来说，这种墙体在内部与外部、城市与乡村之间形成一道隔离线。外国人对中国最强烈的感受，往往就是这种墙体。

在中国的诗歌、历史、故事、书法以及绘画中，垂柳的地位极其重要。垂柳被广泛用来表达对朋友和亲人的眷恋以及思乡情结。其柔软的枝条、纤细飘逸的身姿常常是女性美丽的象征。在隋唐时代，有这样的习俗，当朋友或者亲人分开的时候，折下一条柳枝送给他（她），希望彼此相处的时间更长一些，因为在中文中"柳"与"留"谐音，有"挽留"之意，所以给临别的朋友或亲人送柳条，即希望彼此珍视友情或亲情。除此之外，还有一层含义，人们希望无论走到哪里，都能开始新的生活，因为只要把柳条栽种到土壤里，它就能成活，长成新植株。

充满生机的垂柳与坚硬结实的青砖墙连为一体，表达了"城市与自然和谐共存"的思想。"城市"被一道3米高的简单砖墙所包围，看起来没有出入口。游客从两端的开放拱廊进入"城市"，两端的入口由5道拱廊组成，拱廊立面上全部镶嵌镜片。拱廊穿过1.5米厚的墙体，墙顶种植垂柳，与一系列院落相连。柳枝下垂，呈拱形覆盖着院落，挂有1000个小铜铃，在风的吹动下奏出悦耳的音乐。铜铃的音调与下面院落的宽度相协调。往里走，拱道数量不断增加，游客不得不仔细考虑一下，往哪里走，走哪一条路，就像是穿过一座迷宫，需要不断地选择。与此同时，没有人知道要往哪里去，希望得到什么。这是一种滑稽有趣的感觉，游客带着某种渴望，体会"发现的喜悦"。

院落横向两侧，镜面墙体令人产生一种空间幻觉。游客穿过院落末端，进入黑暗、封闭的出口走廊，镶有单面镜片的墙体面朝花园，中央有一排排呈网格状排列的垂柳，这里似乎是一片不被外界打扰的"世外桃源"。游客一时间不禁凌乱起来："我从'城市'穿越到了'自然'？"

从黑暗盖顶的廊道出来之后，游客会发现，穿过院落时所看到的许多镜子，实际上都是单向镜，通过这些镜片，从隐藏的黑暗廊道中，可以观察他

City and Nature Master Garden

Xi'an, China (2011)

Martha Schwartz Partners was one of nine international landscape design firms to be invited to design a small garden installation on the theme of "The Harmonious Co-existence of Nature and the City" at the 2011 International Horticulture Exhibition in Xi'an, China. According to the guiding ideas of the organization committee, the design scheme should fully consider local building materials and constructional methods. The garden is encouraged to be comfortable, beautiful, and easy of access.

The garden installation is composed of four elements: traditional grey brick walls and paving, weeping willows, one-way mirrors, and bronze bells. The aesthetic direction was derived partly from vernacular Chinese architecture and its close relationship to nature.

Grey brick has long been the principle construction material for vernacular architecture. Such walls are the most popular element to create space and protect privacy in cities, and are frequently used in palaces to express power. For many people living in courtyard houses for generations, these walls provide the separation line of the inside and outside world, city and country. These walls are impressive in the eyes of foreigners.

The weeping willow has a special place in Chinese poems, history, stories, calligraphy, and painting. Weeping willows have been intensively used to express longing, for friends or home, and feelings of nostalgia. They are often used as a symbol for feminine beauty, for its soft, subtle, lissom and graceful figure. During the dynasty of Sui and Tang, when people had to separate with friends or families, they were often given a piece of willow, as a symbol of a lasting and deep affection; through a piece of willow, people hope to start a new life. When the willows are planted in the soil, they will come to life.

The combination of living willows and solid grey walls is an expression of the harmonious co-existence of nature and city. The "city" is entirely walled by simple, 3-meter-high brick walls that seem to have no entrance. One enters the "city" through two ends of an open hallway created by a blank but totally mirrored wall facing a façade of 5 archways. These archways penetrate 1.5-meter-thick walls with weeping willows on top and connect to a series of courtyards. They are overarched with weeping willow branches which are hung with over 1,000 small tuned bronze bell wind chimes. The sonic pitch of the bells is aligned with the width of the courtyards below. The number of possible archways to move through increases as one begins to walk through the space, creating a situation where people must begin to choose where to go and what route to try—an endless choice of routes through the maze. At the same time, no one quite knows where they are going and what to expect. It creates an experience of fun, discovery and perhaps some anxiety.

At each end of the transverse courtyards are mirrored walls which create an illusion of infinite space. As one penetrates the last of these courtyards, one enters a dark, enclosed exit corridor and is confronted with a wall of one-way mirrors facing a mirrored garden room with a grid of willow trees that seems to go on forever. One abruptly transitions from endless city to endless nature.

Exiting via dark covered corridors, one discovers that many of the mirrors they had encountered on the way through the transverse courtyards are actually one-way mirrors, through which they can observe others from the hidden dark corridor. This effect comes at a surprise to the visitors who were not aware until now that they can be watched from behind the mirrors. People can vicariously and secretively watch newcomers in the maze while they hide in the dark corridor.

The garden is a minimalist work of contemporary land art that speaks to the antiquity and timelessness of China, the flexibility and durability of its culture and people. It embodies the harmony and co-existence between Yin and Yang, light and heavy, masculine and feminine. It is rich by its own simplicity. Everybody can sense it in their own way.

| 艺术作品 | Art Commissions | 65 |

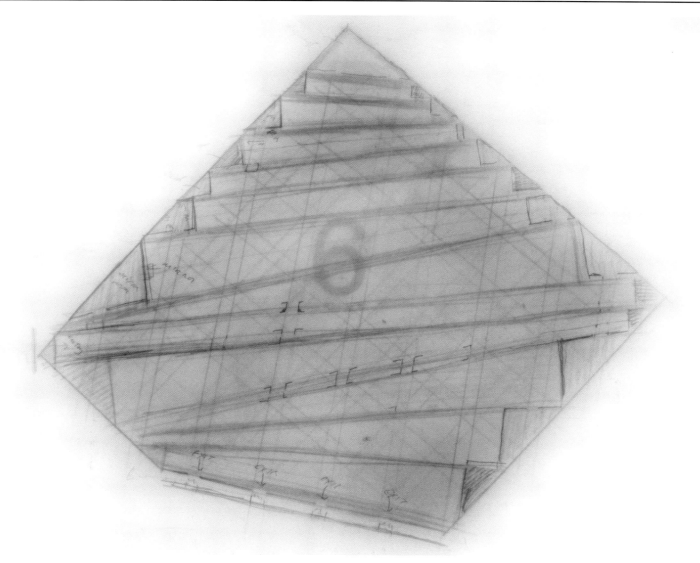

人的一举一动。与此同时，他人也可以躲在镜子后面"窥视"自己的所作所为。抑或在迷宫里，大家躲在黑暗的走廊里，对新来者进行一番秘密观察。

得益于中国古老悠久的历史以及五千多年丰富多彩的民族文化，这个花园堪称当代极简主义的艺术杰作。它体现了阴与阳、轻与重、男性与女性之间的协调与共存。它虽然看起来简单，但却有丰富的内涵。每个人都可以用自己的方式，对它进行一番感受与体验。

艺术作品   **Art Commissions**   69

住宅景观　　　　　　　　　　　　　　　　　　　　　　　**Residential**

住宅景观 Residential

联排住宅

日本福冈，1991

这个项目是一个包括6栋分别由马可·迈克、OMA、史蒂芬·霍尔、奥斯卡·土斯凯特、克里斯蒂安·德·保桑巴克和石山修武设计的公寓大楼组成的多户住宅社区总规划的最后一个阶段。在第一阶段，玛莎·施瓦茨及合伙人设计事务所的设计方案（包括街景、几个院落和入口院落）得以付诸实施。在这个最后阶段，除了项目本身的一些问题需要解决之外，景观设计的主要目的是把项目中不同的组成部分、不同建筑师的设计风格组合成一个整体，并且为项目创造一个令人印象深刻的标志性元素。

由于建筑所处的位置不同，场地设计让一些彼此分离且不规则的剩余空间个性十足。通过对形态、图案和材料的运用，景观小品也独具特色。雕塑般的土丘，穿过竹林，把中央空间整合在一起，并为人们提供各种不同的户外体验。由草坪和岩石组成的鱼丘，在斑驳的冠影和竹林垂直的树干中游动，让人感到神秘莫测。

在这块场地上，为241辆小汽车和450辆自行车提供停车位。在工作日，当小汽车和自行车"缺席"之时，景观小品围合出各种限定性空间，让人们在此开展一系列活动。停车和贮存所应有的秩序井然，通过图形表达出来，使平整的地面和墙体充满活力。主要停车区域被改造成一个棕榈花园；花园中有一片高大的树林，既具有净化空气的功能，又尽显仪式性风采。

Nexus Housing

Fukuoka, Japan (1991)

This project was the final phase of a master plan for a multi-family housing community which includes six apartment buildings designed by Mark Mack, OMA, Steven Holl, Oscar Tusquets, Christian De Portzamparc, and Osamu Ishiyama. In the first phase, designed by Martha Schwartz Partners was implemented for the streetscape and several courtyards and entry courts. In this final phase, aside from resolving programmatic issues, the major design objectives for the landscape were to unify the distinct parts of the project and diverse architectural styles of the participating architects, as well as to give the project a memorable identity.

The implemented site design gives an identity to the disparate and irregularly shaped remnant spaces that resulted from the building locations. The landscape that has been created asserts its own identity through the use of form, pattern, and materials. Sculptural mounds migrate through a Bamboo Forest to unify the central space and provide a variety of outdoor experiences. Grass and stone Fish Mounds swim mysteriously through the filtered canopy and vertical trunks of the bamboo.

The site program requires accommodation of parking for 241 cars and 450 bicycles. Landscape Rooms create a variety of defined spaces that allow for other activities when the cars and bicycles are not present during the workday. The order required by parking and storage are interpreted in the landscape as graphic patterns that activate the ground plane and walls of the Landscape Rooms. The major parking area is transformed into a Palm Court Garden by incorporating a majestic grove of trees into a space that is both functional and ceremonial.

住宅景观   Residential

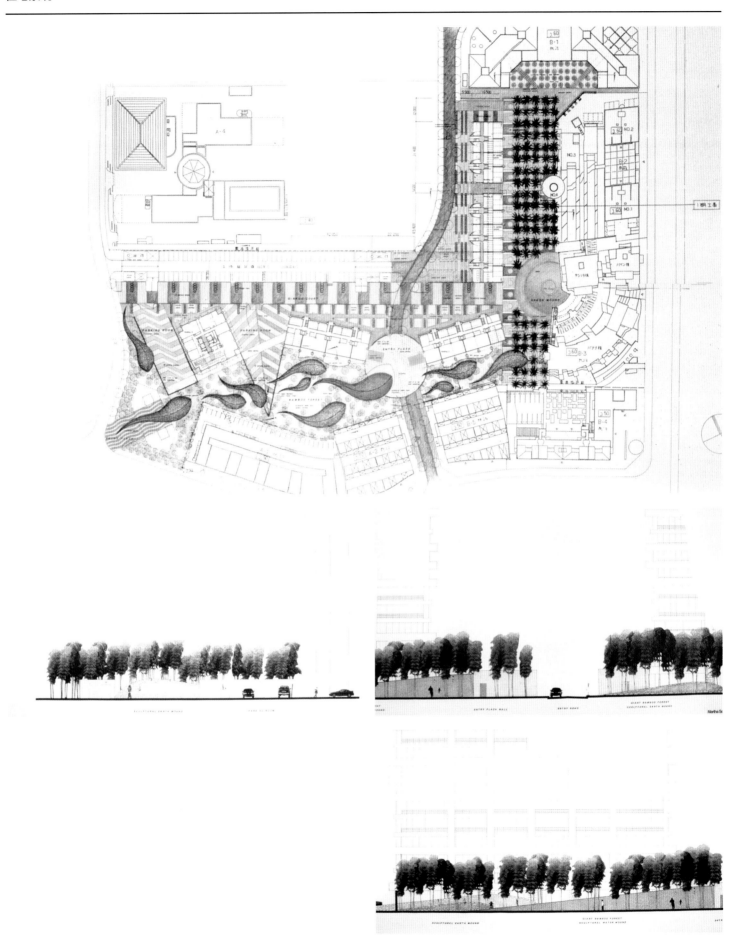

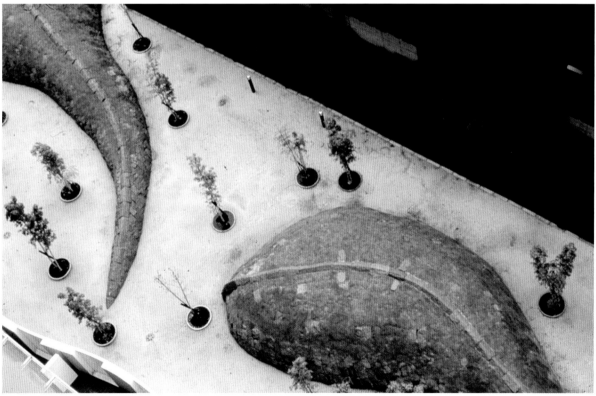

| 住宅景观 | Residential | 78 |

迪肯森住宅

美国新墨西哥州圣达菲，1991

迪肯森住宅坐落在山顶边缘，场地壮观，视野开阔。考虑到这种场地条件，景观设计设置了一系列结构元素，面朝外侧，融入当地景观之中。中央骨干是一条装饰性篱笆，一直延伸至房屋前面，在房屋后面又重新出现，与长排紫叶李并列，穿插在灌木景观之中。游泳池、游泳池平台以及被绿草覆盖的屋顶平台面向开阔的田野，给人留下深刻的印象。

相反，入口景观相对独立且私密。由粉饰灰泥围合成的系列景观空间和廊道一直通向建筑前院。来此参观的游客可以把车停靠在有墙壁的汽车庭院，沿着一道墙体和一排白杨树组成的走廊，抵达入口下沉花园。这里有一片橄榄树林和四个喷泉。地面采用砾石铺装，一系列直角相交的小渠把地面分隔开来。小渠采用色彩鲜艳的瓷砖覆面。这些小水渠与喷泉相连。在每一株树的下面，岩石界定出一系列方形和圆形空间。夜晚，在灯光的照耀下，色彩鲜艳的金属板在喷泉内部里面闪烁着光芒。

Dickenson Residence

Santa Fe, NM, USA (1991)

The Dickenson's adobe house sits atop the brow of a hill and visually commands an expansive site with spectacular views to the horizon in several directions. Responding to these views, the landscape design is organized as a series of gestures oriented outward into the native landscape. A central spine is defined by an ornamental fence which asserts itself at the front of the house and reappears at the back to align with a long line of purple plum trees that intervene in the scrub landscape. A swimming pool, pool terrace, and a turfed roof terrace gesture outward and orient to impressive views.

In contrast, the entry landscape is self-contained and intimate. A series of stucco enclosures creates a sequence of landscape rooms and corridors leading to the front of the house. A visitor to the house parks in a walled motor court, moves down a walled corridor containing a single line of poplar trees, and arrives at a sunken entry garden with an olive grove and four fountains. A gravel panel defines the floor of the room which is divided by an orthogonal series of brightly colored, tile-clad runnels. These runnels interconnect the fountains while rocks define a series of square and circular spaces at the base of each tree. Lit at night, the fountains, with brightly colored interior metal panels, glow from within.

Residential

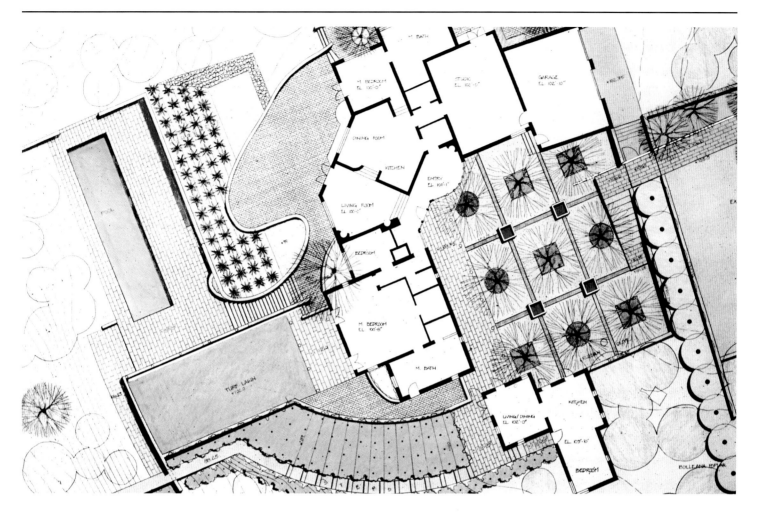

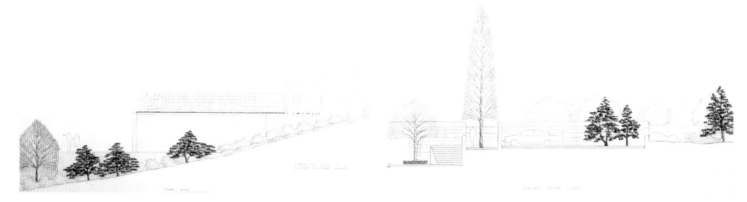

住宅景观 Residential 80

住宅景观　　　　　　　　　　　　　　　　Residential

戴维斯住宅

美国得克萨斯州埃尔帕索，1996

在干旱的环境中，在英式花园中生活25年之后，戴维斯住宅的业主希望做些改变，对花园的一部分进行重新设计，他们希望新花园使用仙人掌类植物，维护强度低，具有墨西哥风格，视觉上看起来与原有的花园区别开来。为了实现业主的愿望，在原有的英式花园中，新建花园用墙体包围起来，看起来有点"笨拙"。这个墨西哥风格的新花园，色彩亮丽，种植着各种仙人掌类植物。

简单的混凝土墙体、喷漆、砾石和仙人掌类植物被用来对带围墙的墨西哥风格花园进行重新阐述。设计方案由一系列盒子组成，园中套园，好像是对住宅风格的模仿。在不同的花园空间之中，单一地种植仙人掌类植物，旨在营造一种盒子内外空间模糊难辨的视觉效果。

Davis Residence

El Paso, TX, USA (1996)

After living with an English-style garden for 25 years in a dry environment, the owners of the Davis residence were ready for something different in the redesign of a portion of their residential setting. They wanted the new garden to be low maintenance with cacti, have a Mexican influence, and be visually separate from the existing garden. To accommodate the owners' wishes, the new garden was developed as a kind of "folly" enclosed by walls within the existing English garden. This new garden of Mexican influence incorporates bright colors and various plantings of cacti.

A simple palette of concrete walls, paint, gravel, and cacti is used to reinterpret a Mexican walled garden. The design is a series of boxes, gardens inside of gardens, presenting a metaphor of the house typology. Singular plantings of cacti inhabit the different garden rooms as the space contained by one box and outside of another becomes intentionally ambiguous.

住宅景观 Residential

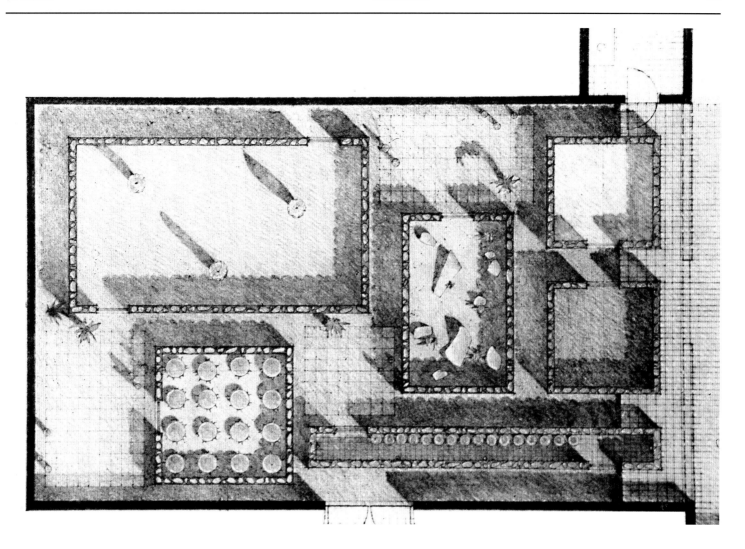

住宅景观 / Residential

日本岐阜北方公寓

日本岐阜，2000

这个住宅庭院项目，是"住房设计女权主义"试验的一部分，包括四座公寓大楼，分别由高桥明子、妹岛和世、克里斯廷·霍利和伊丽莎白·迪勒设计。按照项目的总体规划，庭院处于四座相互独立的公寓之间。因为建筑设计的多元化彰显出鲜明的场地特征和几何形态。景观设计力求将特征鲜明的各个部分统一起来，营造令人难忘的标志性景观。

在被用于住宅之前，这里原先是一片稻田。隆起的堤坝和下沉的稻田所构成的几何形态，暗示在此可以打造一系列下沉花园空间以及景观设施，例如，水景、儿童游乐设施、公共艺术雕塑等，以供人们赏景和休闲娱乐。在柳树庭院的下沉区域中，种植着柳树和湿地植物，并设置了木板步行道。四季花园由四个小花园组成，表现出每个季节的特征，用彩色玻璃包围。在岩石园中，带有踏步石和岩石的圆形喷泉，不定期间隔喷射，形成儿童玩乐池塘。此外，还有一些其他花园空间，包括樱桃前院、鸢尾河、跳舞广场、儿童游乐场、运动场、溪流和竹园等。每个不同的空间都为住户提供了一种不同的体验。

Gifu Kitagata Gardens

Gifu, Japan (2000)

This courtyard project is part of an experiment in "feminism in housing design" which also includes four apartment buildings designed by Akiko Takahashi, Kazuyo Sejima, Christine Hawley, and Elizabeth Diller respectively. In the project master plan, the courtyard is situated between each of the four separate housing blocks. Because of the diversity of architectural design found within the project, strong site imagery and geometry have been created for the courtyard to unify the distinct parts of the project and to give the project a memorable identity.

Before its present use for housing, rice paddies existed on this site. The geometric pattern of raised dikes and sunken paddies provides the metaphor for creating a series of sunken garden "rooms". These rooms offer a variety of opportunities for passive enjoyment or active play including water features, children's play opportunities, and public art. In the Willow Court, a sunken, flooded area with willow trees and wetland vegetation is made accessible by a wooden boardwalk. The Four Seasons Garden is a series of four miniature gardens that capture the spirit of each of the seasons and are enclosed by colored glass walls. In the Stone Garden, a circular fountain with stepping stones and rocks that spit water at irregular intervals creates a children's play pool. The other garden rooms are the Cherry Forecourt, Iris Canal, Dance Floor, Children's Playground, Sports Court, Water Rill, and Bamboo Garden. Each of these rooms provides a different experiential opportunity for the people who live in this community.

住宅景观 Residential

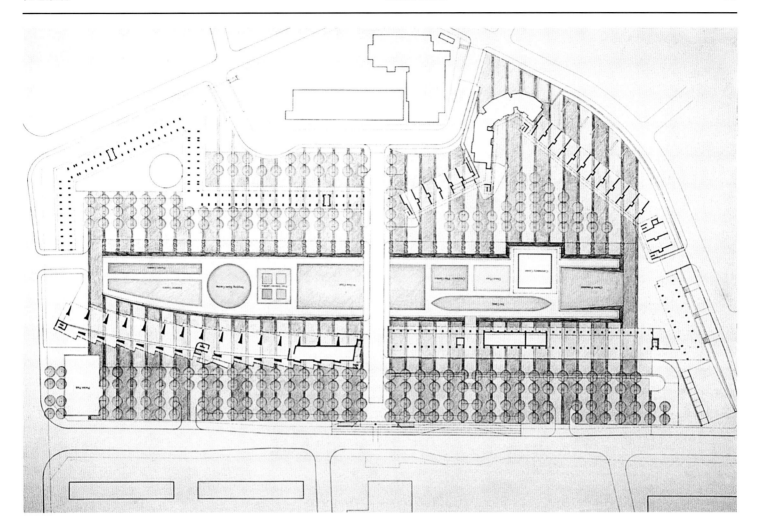

住宅景观 / Residential

保罗林克庭院

德国柏林，2000

这五个庭院的创作灵感源自格林兄弟童话故事。这里原先是一个废弃的工厂，现在要把它改造成带阁楼的豪华公寓。坐落于一个地下车库之上，处于周围高楼的阴影之中，这五个庭院组成了一系列人为创造的神奇画面。为了把传统的童话故事转变成建成环境，设计方案需要与德国的民间传说建立起文化连接。设计方案为这五座庭院打造了令人难忘的标识和富有想象力的景观。庭院周边六层高的建筑遮挡了阳光，童话故事中明亮鲜艳的色调，为庭院带来了生机。充满娱乐色彩的建筑元素、富有想象力的植物种植、雕塑般的地形变化，使每一个花园独具特色。垂直元素使庭院向空中延伸，填充了建筑之间的空白。

其中的一个庭院是根据水精灵的故事设计的。在这个故事中，兄妹二人掉进了一眼井里，被水妖抓获了。孩子们试图逃走，水妖在后面追。为了自卫，男孩扔下一把梳子。梳子魔术般地变成一座山，上面长出数百个牙齿。但是，水妖爬过这座山，继续追赶。女孩扔下一把毛刷，毛刷变成一座山，上面长出数千个钉子。但是，水妖仍然爬过山去。最后，孩子们扔出一面镜子，镜子破碎了，变成一座山，山上满是锋利的玻璃，兄妹二人终于逃脱了。根据这个故事，这个庭院最终成为一片由垂桦构成的密集的小树林。其中的建筑景观元素让人联想起故事中的"魔法山"，供孩子们在其中玩耍。"木梳山"，由体积超大且色彩艳丽的木桩堆积而成，供孩子们攀爬。"毛刷山"，由钢梁构成，上面有很短的彩色尼龙绳，供儿童们玩耍。第三座大山，"镜子山"，黑色橡胶山上布满各种小圆镜，其形象被"扭曲"，组成一个"魔法屋"。"水井"的形态极似一颗蓝色大理石，映衬着下方晶莹的水体。

这五个庭院的设计，基于下列鲜为人知的故事：
1号庭院——六人走遍世界；
2号庭院——十二个兄弟；
3号庭院——女水妖；
4号庭院——桧树；
5号庭院——月亮。

Paul-Lincke-Höfe

Berlin, Germany (2000)

The Grimm Brothers' fairy tales are the inspiration for five courtyard gardens in an abandoned factory being converted into luxury loft apartments. Situated over a parking garage and in the shadow of surrounding buildings, the courtyards are a series of man-made, magical vignettes. By transforming traditional tales into a built environment, a cultural connection is made to German folklore. The design provides a strong and memorable identity for the courtyards, creating a fanciful Ômake-believe' landscape. Because the six-story buildings surrounding the courtyards block the sun, the fairy tale gardens enliven the space with bright, colorful palettes. Each garden's unique character is accentuated by playful architectural elements, imaginative plantings, and sculptural grade changes. Vertical elements extend the gardens skyward, filling the void between the buildings.

One of the gardens is based on a tale about a water spirit. In the fairy tale, a brother and sister fall into a well where they are captured by the water nixie. The children escape, but the water nixie chases them. In desperate defense, the boy throws his comb behind him. The comb magically transforms into a mountain sprouting hundreds of teeth. But the water nixie climbs over it and continues the chase. The girl then tosses her hair brush behind her. The brush transforms into a mountain with thousands of spikes, but the water nixie climbs over it as well. Finally, the children throw back a mirror which shatters and becomes a mountain of sharp glass, allowing them to escape. The courtyard garden corresponding to this fairy tale is a dense, miniature forest of weeping birches. The elements are evocative of the magical mountains in the story, and serve as a playground for children. The "comb" mountain is made of oversized brightly-colored wooden studs stacked for climbing. The "brush" mountain is made of steel beams with short lengths of colored nylon rope attached for children to scale up the mountain. The third mountain, the mirrored mountain, is a black rubber mound covered with small automotive mirrors that distort images, giving a fun-house effect. A "well" is simulated with blue marbles lit from below to represent water.

The designs for the courtyards are based on the following, some lesser known, fairy tales:
Courtyard 1—How Six Made Their Way in the World;
Courtyard 2—The Twelve Brothers;
Courtyard 3—The Water Nixie;
Courtyard 4—The Juniper Tree;
Courtyard 5—The Moon.

住宅景观　　　　　　　　　Residential

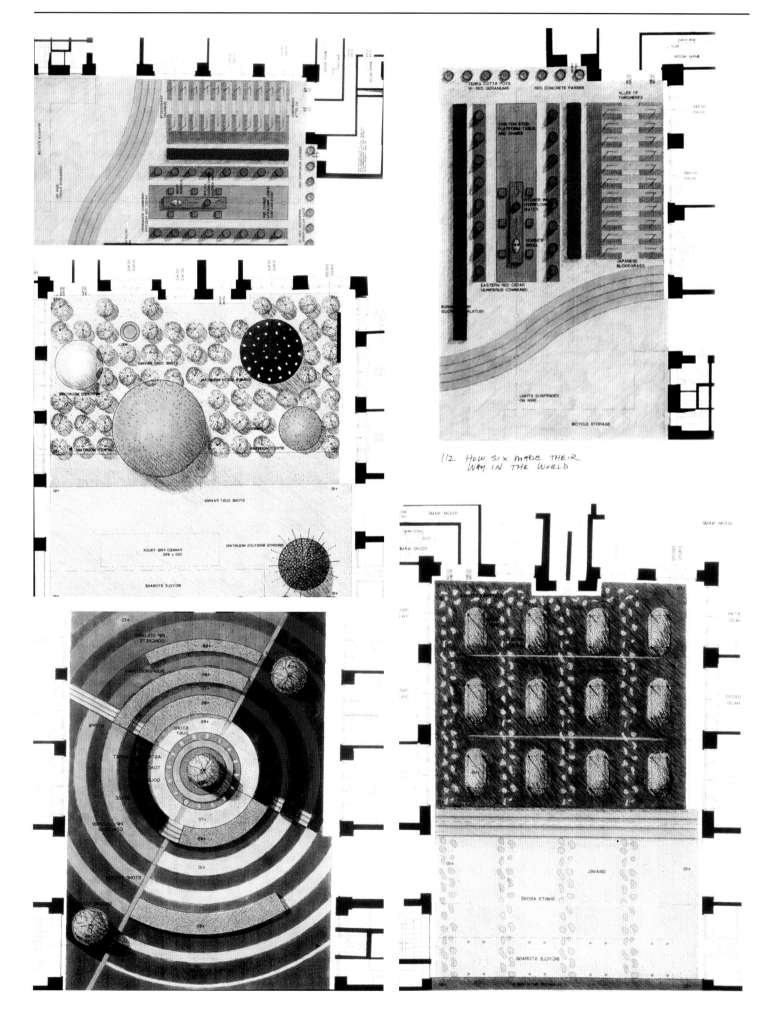

1/2 HOW SIX MADE THEIR WAY IN THE WORLD

住宅景观　　Residential　　104

住宅景观　　　　　　　　　　　　　　　Residential　　　　　　　　　　　　　　　105

住宅景观　　Residential

纳提克努韦勒屋顶花园

美国马萨诸塞州纳提克，2008

纳提克努韦勒屋顶花园力求向人们提供一种奇特的空间和视觉体验：从花园中走过，一览不断变化的景色和丰富多元的色彩；从公寓大楼方向俯视花园，鲜明的图形标识映入眼帘。这个花园位于公寓楼的第6层，在商场的顶楼，处于商场天窗、私密公寓空间与邻里公共空间之间。

在功能上，这个屋顶花园可以作为社交空间，促进邻里之间的交流与互动。蜿蜒曲折的复合木板小路穿过宽广的屋顶，直达座椅休息区，周边环绕着绿篱。落叶树木作为灵活的屏障，提供遮阴，随风摇曳间产生垂直的景观效果。

曲线形的小路环绕着巨大的圆形种植容器。这些容器由耐候钢制成，里面种植装饰性草本植物、开花多年生植物以及常绿灌木。景天属植物所形成的丝带与交替变换的鹅卵石，形成生动活泼的格局，构成小路和种植容器的背景。屋顶花园的两端，小路转换成一系列带有台阶的平台，从连续一致的空间体验转变为在景天属植物和鹅卵石之间的行走探险。在弯曲的步行道上，每隔一定距离，安装小型圆形喷泉，里面是彩色碎玻璃，给人带来惊奇。两个大型圆形结构与种植窗口尺度相仿，作为果岭，构成休闲娱乐空间。此外，还有由岩石组成的大型圆环，与地面商场景观入口所使用的材料产生回声。

这个屋顶花园中有两处向外延伸，为公寓拓展了一些私人领域。这个屋顶花园有助于当地居民建立社交联系、放松休闲，并为其提供了美丽、愉悦的空间和视觉体验。

Nouvelle at Natick

Natick, MA, USA (2008)

The rooftop deck of the residences at the Natick Mall has been designed to be experienced spatially and visually: there are shifting views and rich colors as one walks through and the rooftop as a whole presents a strong and clear graphic identity for those viewing the roof from their apartments above. The deck is located at the sixth floor level of the Residences, above the mall, and in between the mall skylight, residence amenity space, and individual condominium units.

The rooftop functions as a social space, connecting the residences and encouraging interaction between neighbors. A meandering, composite wood path of gracefully curved forms weaves across the expansive roof and filters out into seating areas nestled in hedges. Deciduous trees act as living screens, providing shade and vertical interest as they shift with the wind.

The curves of the path embrace large round planters of corten steel containing a mixture of ornamental grasses, flowering perennials, and evergreen shrubs. A lively pattern of alternating stone and sedum ribbons form a context for the curving path and planters. At the either end of the roof deck, the path transforms into a series of stepping platforms, shifting from the steady experience of the continuous decking into an exploratory walk within the sedum and stones. A series of small circular fountains of colored crushed glass punctuates the walkway for an element of surprise. Two large circles, at the scale of the planters, serve as putting greens, making space for active recreation. Additionally, there are large circles of stonework, echoing the materials used in the landscape entryway of the mall on ground level.

Two extensions of the roof deck allocate private space to apartments at a plus 4-foot level. Together, the rooftop decks provide for social connectivity among the residents, opportunities for relaxation, and an engaging visual experience when viewed from above.

Residential

住宅景观

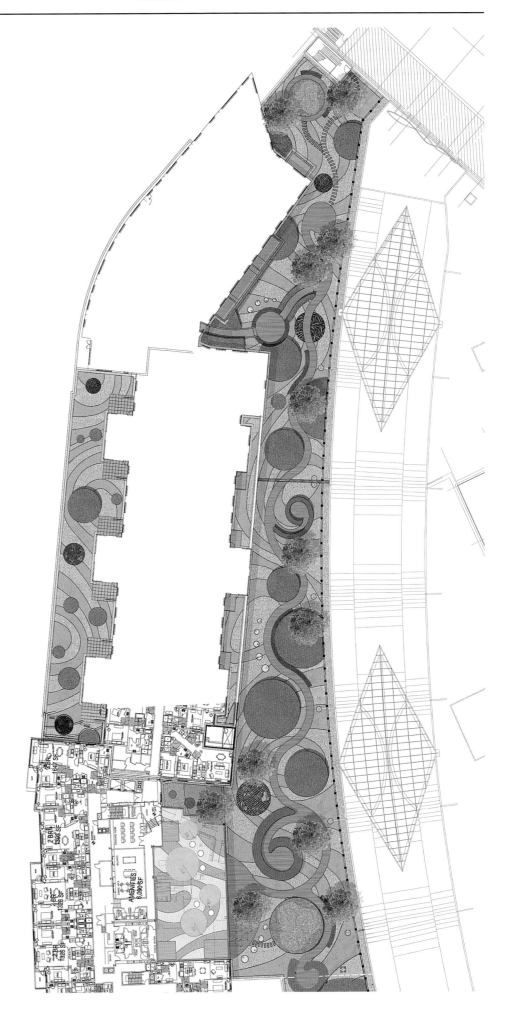

酒店和旅游胜地景观 **Hotels and Resorts**

| 酒店和旅游胜地景观 | Hotels and Resorts |

德拉诺酒店

美国佛罗里达州迈阿密海滩，1995

德拉诺宾馆的设计简洁明快、干净清爽，让人感到舒适惬意，为迈阿密海滩带来些许生机。酒店景观以及经过改造更新的基础设施浸润于这片开阔的海滩之中。宽大的草坪上，9.1米高的华盛顿棕榈树构成花园的骨架，其中还有由柑橘树构成的小树林。游客可以在凉爽且有遮阴的躺椅上放松休息，欣赏银光闪烁的蓝色游泳池，游泳池贯穿整个场地。延伸的花园和游泳池，通过一个具有雕塑感又显眼的楼梯与酒店高层露台相连，由草坪与铺装材料构成的踏面相互交织。台面上，按照一定的几何规则，摆放着一些种植开花植物的陶罐。花盆中种植开花植物。植株溢满花盆，并略微下垂，对连接建筑、平台和花园的通道起到装饰作用。这条通道一直延伸至游泳池，面朝大海。

Delano Hotel

Miami Beach, FL, USA (1995)

Delano South Beach Hotel with its simple, crisp, clean and modern sense of ease reinvented South Beach hotel design. The landscape associated with the renovations to this existing hotel complement and reinforce the playful, elegant and chic public spaces that permeate this luxury abode. The garden is structured by 30-foot tall Washingtonia Palm Trees on a grand lawn carpet interplanted with a citrus grove under which guests can relax in the cool shaded recesses as they look out towards the glimmering blue swimming pool that runs the length of the site to the beach. The stretch of garden and pool are connected to the upper level hotel patio via a strong and sculptural staircase that interlaces the lawn with paved material of the treads. Geometrically arranged on the steps are placed a series of large clay pots brimming over with flowering, draping plants that ornament this passage and link between building, patio and garden below that advances to the pool and then the sea.

Hotels and Resorts

酒店和旅游胜地景观 | Hotels and Resorts | 115

酒店和旅游胜地景观　　　　　　　　　　　　　Hotels and Resorts

迪斯尼乐园东入口广场

美国加利福尼亚州阿纳海姆，1998

迪斯尼乐园东入口广场由三个不同的区域组成：中心广场、有轨电车东区（超级高速公路区）、有轨电车西区。在特征和功能上，这三个区域有明显的不同。尽管外在形态有所不同，但是，它们构成一个空间序列，把各种舞台事件有序地串联起来。受古典巴洛克风格花园的启发，这三个不同的区域采用线性方式有序组织，各种不同的事件和空间，通过线性人行通道相连。这条通道使整个场地变得连贯有序。植物种植和其他元素的空间组织，运用几何形体和韵律，把各个不同的部分"捆绑"在一起，创造出某种连续性。

在机动车道和步行区之间，有轨电车东区（超级高速公路区）形成了明显的边界。在"高速公路"上，一系列图案和标志被夸张和重复地使用，比如，人行通道、圆锥形隔离墩、路灯等，令原本单调乏味的"高速公路"变得情趣盎然。游客进入超级高速公路区之中，通过一系列生动形象的视觉语汇，"高速公路"变得更加开阔且色彩鲜明，为游客提供了一种愉悦、畅快的空间体验。

从宾馆区开来的公共汽车，沿着线性交通岛停下来载客。人行通道体量夸张，引人注目，使耀眼的交通岛醒目，对人行通道起到明显的强化作用。交通岛周围的标准高速公路照明灯具呈规则式排列。这种照明灯具具有金属般的色彩装饰效果，不同的交通岛表现出不同的色彩。夜晚，彩色照明明亮且绚烂，有助于游客寻找道路和辨别方向。

路桩的形状类似于圆锥形的隔离墩，体量巨大，被漆成绿色，就像欧洲古典花园中经过整形与修剪的树木。由路桩组成的线条蜿蜒地穿过整个东广场，彼此间隔1.5米，之间有鲜黄色的警戒线。瓷砖在行人和机动车之间产生触觉和视觉上的"分离"之感。

Disneyland East Esplanade

Anaheim, CA, USA (1998)

The Theme Park Entry Esplanade project is composed of three distinct areas: the Central Plaza Area, the East Tram Area or "Hyper-Highway", and the West Tram Area. All three spaces are distinct in terms of character and function; however, they are conceived as a series of spaces which, although different in expression, form a choreographed sequence of events. The three distinct spaces are organized in a linear fashion, inspired by gardens of the classical Baroque Period, where episodic events or spaces are joined through a linear progression of the pedestrian through the spaces. This journey, or pathway, links and brings coherence to the site. The use of geometry and its rhythm through the plantings and elements create a continuity which ties all parts of the design together.

The "Hyper-Highway" concept was conceived as a tool to create clear boundaries between vehicular and pedestrian areas. Exaggeration and repetition of highway motifs, such as pedestrian crosswalks, traffic cones, and highway lights are used not only to help create clear and defined spaces, but also to create fantastical places. As a visitor turns into the environment of the "Hyper-Highway", he or she will experience the world as a bright and more vivid place, as the visual language of the highway becomes bigger, bolder, and more colorful.

Shuttle buses from hotels in the area drop off and pick up passengers along linear traffic islands. These glorified traffic islands are marked by an insistent and oversized pattern of "crosswalk" to indicate clear, strong pedestrian access to the entry. Each traffic island is flanked by formal rows of highway lighting standards. The light standards have a metallic color finish. They glow a different color in each island. The brightly colored light provides not only a dazzling field of glowing colors at night, but also aids in way-finding and orientation.

The bollards, shaped as oversized traffic cones, painted green, allude to clipped topiaries in the European Garden Tradition. A single line of bollards winds through the entire East Esplanade on a 5-foot-wide strip of bright yellow "detectable warning" tile. The tile provides the ultimate visual and tactile separation between pedestrians and vehicles.

Hotels and Resorts

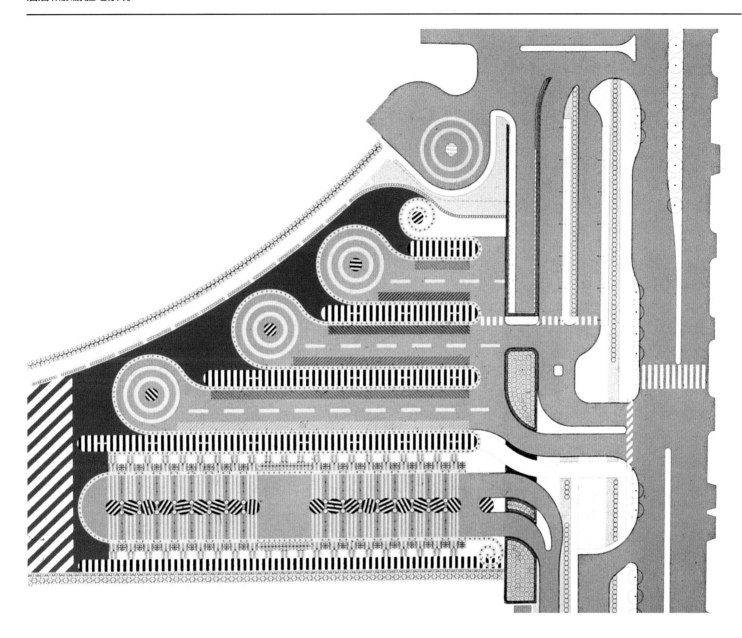

Hotels and Resorts

酒店和旅游胜地景观　　　　　　　　　　　　　　Hotels and Resorts　　　　　　　　　　　　　　119

商业景观　　　　　　　　　　　　　　　　　　　　　　　　**Commercial**

商业景观	Commercial

里约购物中心

美国佐治亚州亚特兰大，1989

亚特兰大市中心的里约购物中心庭院中，摆放着呈阵列放置的镀金青蛙，它们都好像在对一个几何球体表示着尊敬。这个由 ARQ 设计的里约购物中心，可以大胆地宣称它正处于繁忙的十字路口中，等待着复兴。这个球形处于购物中心的视觉中心，购物中心的第一层商店所面向的院落低于街道 3 米。

草坪、地面铺装、岩石、建筑相互重叠，构成空间的主体框架。广场通过各种几何体分层构建，包括直线、圆、球体和立方体。这些元素在神秘的黑色池塘中相遇，直线光纤在夜晚跳跃、闪烁。一条漂浮的通道由其上方所连接的桥梁反射而成，把购物区与另一边串联起来。

青蛙在球体的下方呈网格状排列。距离地面 12.2 米高的球体坐落于连接道路和庭院的斜坡上。由青蛙组成的网格沿着斜坡继续向下延伸，穿过池塘，面朝这个巨大的球体，就像在参拜。地球仪为藤本植物提供支撑，并配有喷雾喷泉。除了这个视觉焦点之外，有一个广场，可以作为人们的聚会场所；它好像一个圆形的酒吧，直抵屋顶的竹林。此外，还有一间视频工作室，由艺术家达拉·伯恩鲍姆设计并搭建。

Rio Shopping Center

Atlanta, GA, USA (1989)

A squadron of gilded frogs worships a geodesic globe in the courtyard of a specialty shopping center in midtown Atlanta. With architecture designed by Miami-based Arquitectonica International, Inc., Rio Shopping Center boldly asserts itself among the chaos of a cluttered intersection in an area ripe for revitalization. The globe serves as a beacon for the retail center whose first level of shops opens onto a courtyard 10 feet below the street.

Overlapping squares of lawn, paving, stones, and architecture form the basis of the design. The squares are layered with other geometric pieces—lines, circles, spheres, and cubes. These elements meet in a mysterious black pool which is striated by lines of fiber optics that glow at night. A floating path, reflected above by an architectural bridge, connects one side of the shopping area to the other.

The frogs are set in a grid at the base of the 40-foot-high globe which is located on a slope connecting the road to the courtyard. Alternating stripes of riprap and grass cover the slope. The grid of frogs continues down the slope and through the pool, all facing the giant sphere as if paying homage. The globe, which also provides support for vines, houses a mist fountain. A square plaza beyond this focal point forms a meeting place which includes a circular bar, a bamboo grove that punctures the roof, and a video installation by artist Darra Birnbaum.

商业景观 **Commercial** 127

商业景观 | Commercial

城堡购物中心

美国加利福尼亚州考莫斯，1991

城堡购物中心的所在地，从前是尤尼轮胎和橡胶厂，现在引起了南加州人的注意。那家工厂建于20世纪20年代。装饰亚述寺庙和浮雕前墙由综合应用开发商采取保护措施。景观设计力求保护这道墙体，并提出与周边环境相协调的设计方案。一方面，保持这道墙体的神秘性；另一方面，对于零售商、办公人员、宾馆酒店服务人员等具有较强的吸引力。

这道历史性墙体的中部，靠近通灵塔寺的地方，已经断裂了。在图案丰富的广场上，种植成排的海枣树，形成一个绿洲。建筑沿着墙体两侧延伸，在终端处，有一个小型中心广场，宽45.7米，长213.4米，可以作为人们聚会或举行典礼活动之所。经过特别设计的轮胎形状的环形结构，围绕着每一株海枣树，把步行空间和机动车空间分隔开来。地面铺装采用彩色长方形混凝土地砖，呈棋盘状。在植被和其他元素的衬托下，广场在视觉上有一种下滑的感觉。宾馆和零售院落的入口位于交叉轴线上，两条轴线由建筑和海枣树构成。零售院落就像一个集市，有步行小路和树木遮阴的空间，有遮阳篷和水。所有这些元素和设计理念综合起来，营造了奇妙的环境氛围，让人联想起另一个时代和另一个国度。由开花树木构成的规整的机动车道把中央空间与规划中的宾馆串联起来。从广场到宾馆汽车庭院，地面铺装采用特殊材料。停车场的景观设计让人们联想起南加州和地中海地区的农业景观。干燥、灰色的橄榄树，成排种植，与海枣树绿洲形成鲜明的对比。

Citadel Shopping Center

Commerce, CA, USA (1991)

The Citadel site in City of Commerce, CA, formerly the Uniroyal Tire and Rubber Plant, has captured the imagination of generations of Southern Californians. The factory was built in the 1920's. The decorated Assyrian temple and bas relief front walls were preserved by the developer of this mixed use site. The landscape design needed to retain the wall and create a design in a compatible context. Additionally, it was important to maintain the mystery of the wall while at the same time creating a design that would attract users to the retail, office, and hotel buildings.

At the center of the historic wall, adjacent to the Ziggurat temple, the wall has been breached to reveal an oasis of date palms aligned in rows on a patterned plaza. The buildings are located along two sides and at the terminus of the 150-foot-wide by 700-foot-long central plaza evoking a civic or ceremonial space. Pedestrian and vehicular spaces are separated by specially designed tire-shaped rings which surround each of the palm trees. The checkerboard paving is composed of a series of colored concrete rectangular pavers and visually slides under the plantings and other plaza features. The hotel and entrance to the retail court are on a cross axis framed by buildings and palms. The retail court recreates a Middle Eastern Bazaar—a space of shade trees and paths, awnings and water. All of these elements and the design concept create an environment evoking the mystery of another time and place. A formal allée of flowering trees connects the central space to the planned hotel. Special pavement links the plaza to the hotel motor court. Parking for the project is designed to recall the agricultural groves of Southern California and the Mediterranean. Row plantings of dry, greyish olive trees contrast dramatically against the green palm oasis.

医院景观

Health Care

| 医院景观 | Health Care | 136 |

维也纳北方医院景观

奥地利维也纳，2008

维也纳北方医院景观总体规划包括三个主要部分：康复公园、公共城市广场、机动车临时停车区和急诊服务区。在广场内，平行线性铺装格局把各个活动区域无缝连接起来，涵盖主入口广场、户外咖啡厅、火车站及附属小广场、宾馆及户外会议空间，以及其他循环空间。这种铺装图案对医院主入口前方的空间起到有效的混合和分隔作用，有助于引导行人通行。

乍一看，这座医院建筑就像一个公园，公园式的景观设计使内、外边界有机结合。一条线性体系从里侧的医院建筑向外延伸至外侧的医疗花园、草地以及水体等，在患者治疗过程中扮演着重要角色。在设计方案中，水体是重要的连接元素。安静平和的结构性下沉式室内花园围绕着游泳池，构成一系列湿地，在场地降水管理体系中具有重要作用。

水体两侧种植开花草本植物，并设有草药花园和医疗花园，以丰富景观的视觉效果。花园之间是整齐修建的草坪和潘诺尼草坪。

条带的中央有四座温室，里面种植柑橘，作为冬园。与水景一样，这些冬园把医院空间和花园空间串联在一起，并与医院前庭共同构成患者与医务人员的主要活动空间。

这个设计方案获得了竞赛一等奖，目前正在建设之中，预计于2017年完成。

Vienna North Hospital

Vienna, Austria (2008)

The landscape master plan for Vienna North Hospital is composed of three primary areas: a restorative park, a public urban plaza, and a vehicular drop off and emergency service area. A parallel linear paving pattern seamlessly connects the variety of program areas within the plaza: a main entry sub-plaza, outdoor cafes, a train station and associated sub-plaza, hotel and conference outdoor spill out spaces, and other circulation spaces. The pattern significantly bends and splits in front of the hospital's main entry, helping to direct pedestrian traffic.

The park design blends the boundary between interior and exterior with a linear system that extends from inside the hospital building into a series of therapeutic gardens, lawns, meadows, and water bodies that serve a vital role in the patient healing processes. Water is the primary connective element of the proposal as architecturally defined, serene, sunken interior gardens become a series of wetland surrounded pools that are an essential part of the site's storm water management system.

On either side of the water's riparian edges, flowering meadows, odiferous herb gardens, and therapeutic gardens heighten one's sensory experience of the landscape. Native Pannonian meadows and mown lawns are wide landscape strips between these gardens.

Four glass house structures in the middle of these strips are citrus filled winter gardens. Like the water, these winter gardens mark a significant connection between the hospital interior and garden exterior. They are positive volumes that are aligned with the hospital's atria voids.

The competition entry received the first prize and the project is currently under construction with a completion date in 2017.

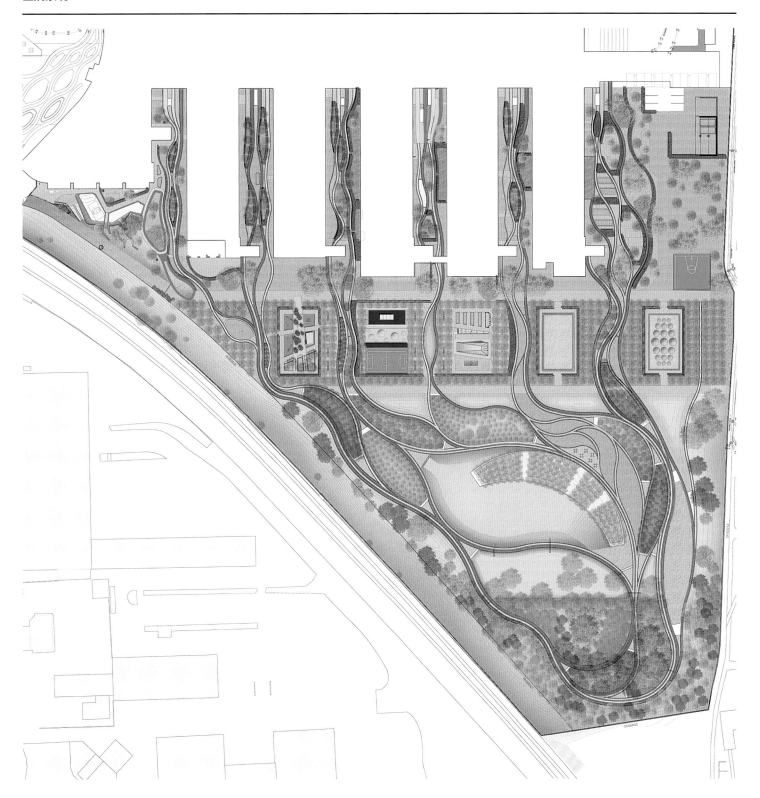

Health Care

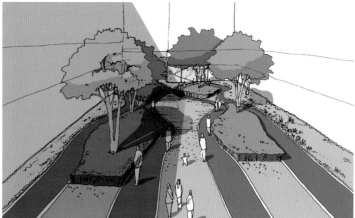

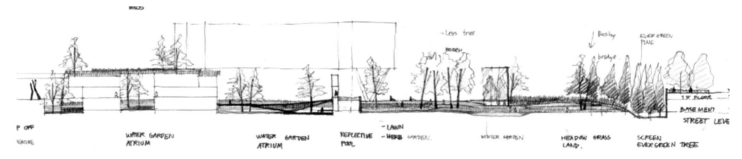
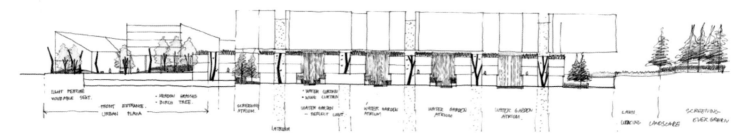

卢森堡南部医院景观

卢森堡阿尔泽特/埃施,2015

这个项目的宗旨是打造完整统一的标志性医院景观。在这个区域中,有农田、工厂和郊区居民。基本的建筑单元呈"菱形"。景观设计通过地形改造与建筑相适应,建筑主体呈台阶式下降的趋势。"菱形"的建筑造型在地表形成连续且流动的格局,道路系统延伸至与医院相邻的场地,与未来的开发区域相连。在项目范围之内,医院建筑和台阶式地形的周边种满各种各样的树木和其他植被,营造了休闲平和的氛围。病人、医生、学生和来访者在这里互动与交流。

场地设计的主要特征是运用多种方法加强降水的排放和收集。市政工程师在初步计算后证明,场地中的洪水可以通过双重可持续性战略加以缓解。

场地硬质地面上的降水,作为战略的第一个层次,排放到重新改造后的现有运河之中,这条运河正好从场地之中流过。此外,设计方案对运河进行重新塑造,并把水体要素塑造成曲线形的湿地,使其一年四季充满生机和活力;根据洪水淹没情况或者正常土壤情况,种植草本植物、树木和灌木;对运河中的植物进行一定的修复。场地中的降水流入运河,有助于消除污染物。经过处理的干净的河水流入场地的东端,与原来的运河重新汇合。或者,如果可行或者有需要,可以让河水发生溢流,流入东侧场地外面的人工蓄水池中。

战略的第二个层次,是屋顶。据工程师估计,屋顶降水的40%,可以被绿色屋顶所吸收。剩下的60%,可以被每座医院建筑附近的地下蓄水池收集起来,用于景观植物的灌溉,实现循环利用。

这个设计方案获得了竞赛一等奖,目前正在规划之中,预计于2019年完成。

Sudspidol Luxemburg

Esch/Alzette, Luxemburg (2015)

The hospital setting has been designed to stand as a unified and iconic presence that integrates within the context of an urbanizing landscape consisting of farmland, factory and suburban residences. The basic building unit presented itself as a "lozenge" shape that the landscape responds to with land forms that populate the site and terrace down from the building masses. The lozenge shape allows for a fluid, continually flowing circulation pattern at ground level within which a system of paths can be extended into sites adjacent to the hospital that link it with future development. Within the project boundaries, the hospital buildings and their terraced land forms, set up a topographically exciting ground plane for a parkland environment, full of trees and vegetation that will be conducive to interaction between patients, doctors, students and visitors providing a restful and calming atmosphere.

A major feature of the site design is the comprehensive approach that has been taken to address storm water drainage and collection. Preliminary calculations by civil engineers determined that flooding on the site can be mitigated by a dual part sustainable strategy.

Storm water that falls upon the site hardscape is one layer that will be drained towards a reconfigured existing canal that currently passes through the site. The proposal reshapes the canal in a soft-engineered way to sculpt this water element into a curvilinear wetland feature that is attractive all year round, planted with grasses, trees and shrubs that can take inundation or normal soil conditions. The plants in the canal will perform phytoremediation, removing pollutants, upon the site storm water that passes through them. The cleansed water will exit the east end of the site where it will rejoin with the existing canal. Or, it can overflow into a carefully landscaped man-made retention basin located outside the project site to the east if feasible and desired.

The second layer of storm water drainage strategy involves the roofs. Engineers estimated that about 40% of roof water will be absorbed by the proposed green roof designs. The remaining, or other 60%, will be collected in underground tanks located near each hospital building and recycled for irrigation of the landscape plants.

The competition entry received the first prize and the project is currently in planning with a completion date in 2019.

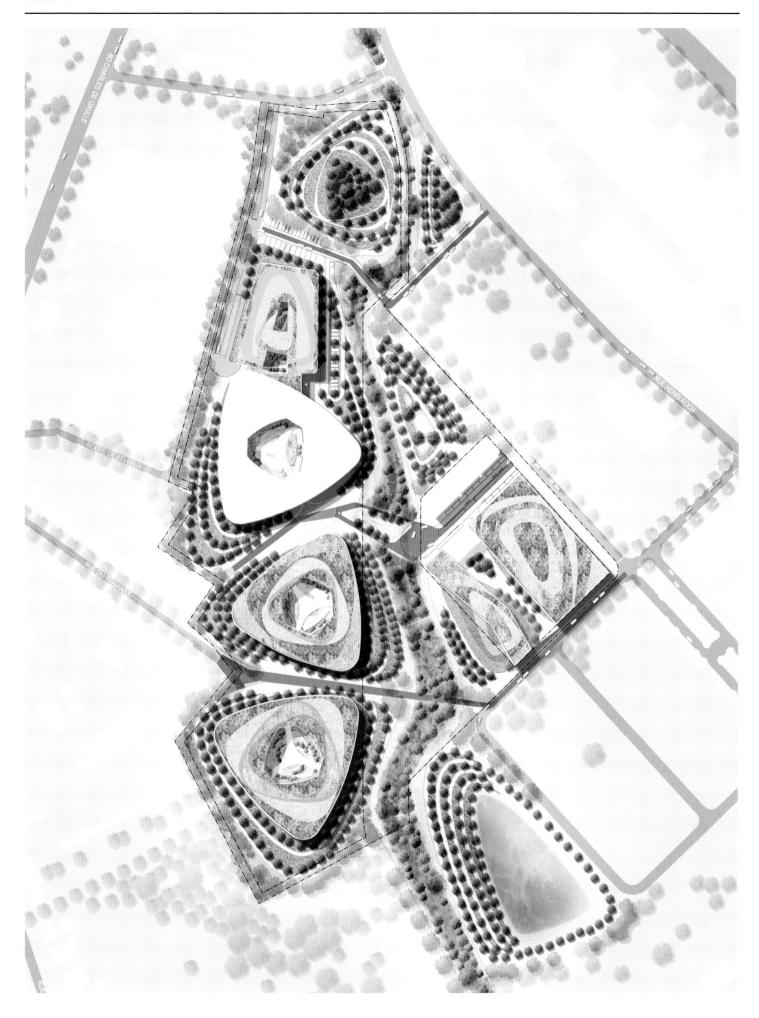

市政景观 **Civic**

HUD广场

美国华盛顿特区,1996

HUD(华盛顿特区住房和城市发展部)广场由马歇·布劳耶于1968年设计。尽管广场前方建筑的外立面线条清晰多变,但是2.4公顷的广场确实是为了符合现代美学而存在的牺牲品。与其设计初衷大相径庭是因为其四周缺乏必需的绿化或者公共设施,而失去了4800名员工的青睐。同时建筑底部由黑色石块构成的坚硬墙体割断了建筑内外的视觉联系。广场设计的目的是突出这座建筑,使其更好地被工作在其中的员工所使用,同时激活整个广场,为人们营造良好的娱乐休闲空间。

广场设计采用圆形图案,由白色、黄色和灰色等不同的色彩构成,让人不禁联想到布劳耶在窗户、墙体和天花板上所采用的几何形体。新广场主要由具有极具冲击的大片平地以及点缀其中的大型植被草盆以及环状白色顶棚所组成。大型草盆容器直径9.1米,双面都可以提供座位。顶棚主要材料为乙烯基塑料,并由钢柱支撑被固定在离地面4.3米的位置。这些顶棚和大型植被草盆所带给人的轻浮的感觉,与沉重严峻的建筑形成鲜明的对比。因为广场是建在一个地下停车场的上方,由于缺乏种树必要的土壤,这些顶棚也能起到遮阴的作用。

广场的灯光设计增强了HUD plaza的辨识度。这些顶棚采用内部照明,夜晚闪闪发光,让人联想起日本公园里的古朴的灯笼。大型草盆的下部由光纤管打出彩色灯光,使它们看起来就像漂浮在由灯光构成的云层之上。安装在建筑底部黑色墙体中的背部照明灯,照映着整幢建筑,也映衬着忙碌于其中的人们,为HUD广场营造了迷人的背景。

HUD Plaza

Washington, D.C., USA (1996)

Although Marcel Breuer's 1968 building for the Department of Housing and Urban Development (HUD) in Washington, D.C. bears a richly textured facade, its 6-acre plaza is clearly a casualty of the Modernist aesthetic. Without trees or public amenities, the plaza was designed to showcase the building, but is virtually unusable by HUD's 4,800 employees. Adding to the desolation of this landscape is the fact that the base of the building is a solid wall of dark stone that prohibits a visual connection between the life of the building within and that without. HUD's objective for the plaza was to reactivate it by commissioning a new design that would also express the agency's mission of creating habitable spaces for people.

The scheme developed for the plaza repeats a circular motif in white, yellow, and grey recalling Breuer's use of geometric designs for screens, walls, and ceilings. The plaza is transformed through a strong ground plane, a series of concrete planters containing grass, and white lifesaver-shaped canopies. The 30-foot diameter planters double as seating. The canopies, fabricated of vinyl-coated plastic fabric, are raised 14 feet above the ground plane on steel poles. In sharp contrast to the heaviness and somberness of the architecture, these canopies and planters appear to float. As this plaza is built over an underground garage, the canopies also provide shade on a plaza that was not designed to support the soil required for trees.

Lighting gives identity to the plaza as well. Lit from within, the canopies glow at night, recalling the lanterns that illuminate paths in Japanese gardens. A fiber-optic tube casts colored light under the planters making them appear to float on a cloud of light. For the dark wall at the base of the building, a backlit mural has been planned to reflect the people and faces of HUD and create a dramatic backdrop for the plaza.

Civic

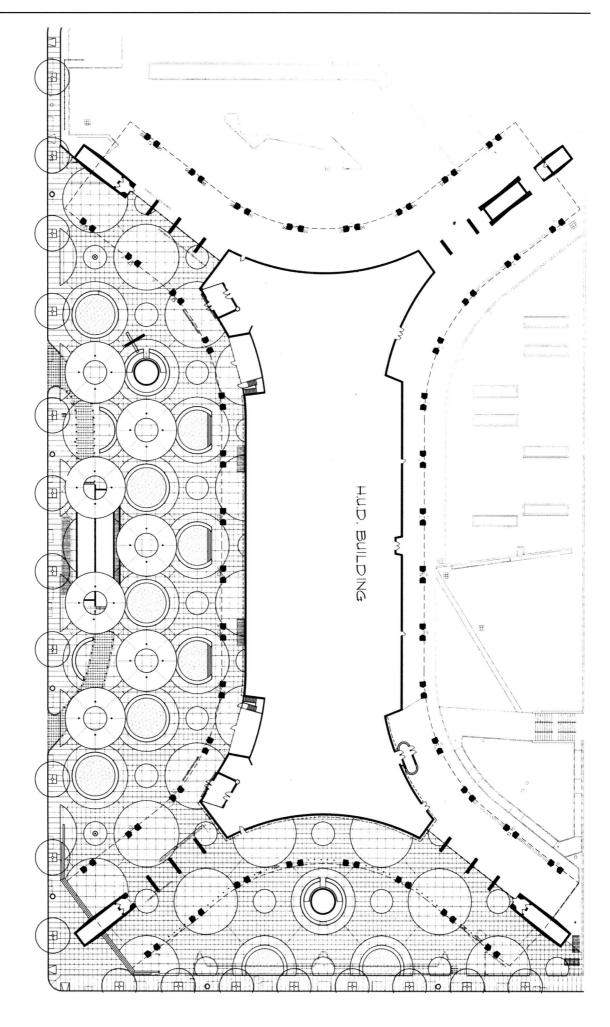

市政景观　　Civic　　150

市政景观 Civic

明尼阿波利斯联邦法院广场

美国明尼苏达州明尼阿波利斯，1996

这个 4645 平方米的广场位于明尼阿波利斯市中心，面朝市政厅。在新建联邦法院的前方，这座法院大楼由 KPF 建筑设计事务所设计。这个项目需要建设一个广场，拥有独特的形象和场所感供市政当局和个人使用。

这个广场完全建立在一个地下停车库上面，体现了明尼苏达州的文化和历史。土丘和原木极具历史意义，成为雕塑般的广场构件。在广场内，这些元素象征着人们根据自己的需求对自然景观的改造和应用。

土丘，是为了唤起人们对地形和文化形态的回忆。这些元素，让人联想到明尼苏达州的冰川和别具一格的丘陵。这些山脉和丘陵凹地可以进行双重解读。在高度 0.9～2.7 米且形状类似于眼泪的土丘上种植着大量短叶松，这是一种矮小敦实的松树，是明尼苏达州北方针叶林中常见的先锋树种。

由原木制成的座凳诉说着类似的故事，让人联想起大面积的森林，吸引移民的到来，成为地方经济基础。木材与明尼苏达州相关联，触及明尼苏达人回忆的核心，给前来广场的游客打上深刻的情感烙印。设计中所采用的明显的导向标志有助于行人穿过广场抵达法院大楼。线性条带状的铺装格局引导行人进入大厅。土丘同样具有导向作用，引导行人进入前门。

明尼苏达州显著的季节变化也在这个广场中反映出来。每年春天和夏季，土丘上生长着多年生植物。有些土丘上覆盖着白色水仙，有些则种植着蓝色欧洲绵枣，形成条带，对地面铺装起强化作用。冬季，明尼苏达州的大雪使土丘的雕塑效果更加突出。

Minneapolis Courthouse Plaza

Minneapolis, MN, USA (1996)

This 50,000-square-foot plaza is located in Minneapolis's civic center, facing City Hall and in front of a new federal courthouse designed by Kohn Pedersen Fox Associates. The program required a plaza designed for both civic and individual activities, with its own imagery and sense of place.

The design developed for this plaza, entirely built on a garage roof, refers to Minnesota's cultural and natural history. Earth mounds and logs, elements of that history, are the plaza's symbolic and sculptural elements. Within the plaza, these components symbolize both the natural landscape and man's manipu-lation of the landscape for his own purposes.

The mounds are intended to evoke a memory of geological and cultural forms. They suggest a Minnesotan field of glacial drumlins, a stylized hill region, or, like a Japanese garden, a landscape that allows a dual reading of scale—a range of mountains or a low field of mounds. Ranging in height from three to nine feet, the tear-shaped mounds are planted with jack pine, a small, stunted, pioneer species common in Minnesota's boreal forest.

The log benches, evocative of the great timber forests that attracted immigrants and provided the basis for the local economy, tell a similar story. The association of timber with Minnesota speaks to the heart of Minnesotans' collective memory, and the plaza leaves a strong emotional imprint on the people who visit it. Strong signals in the design help pedestrians move through the plaza to the courthouse building. The linearity of the striped paving pattern guides the pedestrian into the lobby. The drumlins themselves also provide a directionality to the front door.

Minnesota's strong change of seasons is reflected in the plaza. Each spring and summer, the mounds come alive with perennials. Some mounds are blanketed with white narcissus while others reinforce the paving with stripes of blue scillas. In winter, Minnesota's heavy snows heighten the sculptural effect of the drumlins.

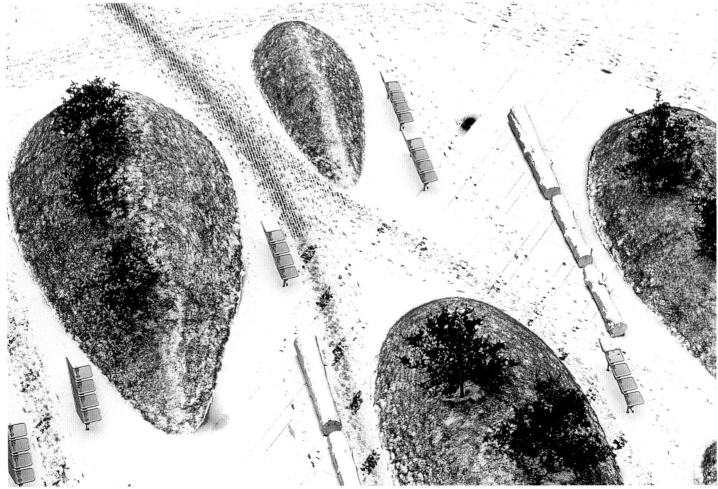

市政景观　　　　　　　　　　　　　　　　　　Civic　　　　　　　　　　　　　　　　　　154

雅各布·贾维茨广场

美国纽约州纽约,1997

1992年,联邦政府出资对雅各布·贾维茨广场地下停车库防水部分进行了维修。防水设施建设期间,广场被破坏,但却为它的复活带来了机会。原来的广场中有理查德·塞拉的雕塑"倾斜弧面",高4.3米。这座雕塑在视觉和实体上均对行为构成阻碍。在它被移除之后,广场空空荡荡的,与周边环境失去了连接。对这个广场进行重新设计的目的是在城市的心脏地带打造一个可供人们使用且生动活泼的开放空间。为了实现这种转变,景观设计师需要运用艺术设计和景观设计的整套设计技术。

新设计的广场与周边环境建立了密切的联系,大量座椅可供人们在这里午餐,或者只是简单地观赏过往行人。广场西北角和东南角的大型种植容器被拆掉了。占据广场大部分阳光地带且长期空闲的喷泉也被移除了。广场与街道实现了无缝衔接。路过广场的人们既可在阳光之下享受阳光,也可在树荫之下享受阴凉。

座椅由纽约市公园中的长凳捻股而成。双股背对背长凳,前后摆动形成环路,进而成为各种形式的座椅。向内的圆环适合小团体活动;外向曲线适合那些独享午餐的人。这些形态多元的鲜绿色长凳使这个平面广场充满活力,就像法国的规则式花园中的刺绣花坛,其中点缀着经过整形修剪的地被植物,边缘是树木和建筑。鲜绿色的长凳使这个大多数情况下处于遮阴状态下的广场更加充满活力。

在广场中,座椅沿着"整形植物"和1.8米高的半球形草坪呈螺旋状盘旋。在天气炎热之时,草坪中央向外喷雾,同时,广场向人们提供各种熟悉的午餐时段景观用具,比如,蓝色搪瓷饮水喷泉、橘黄色网眼垃圾桶以及标准照明用具等。这些用具都是从纽约奥姆斯特德传统公园中借用的,但是,与它们的历史先驱相比,每一种用具又多少有点改进。在纽约市,这些用具堪称景观小品,创造了独特的景观文化艺术,在景观设计界颇有名气。纽约作为多种文化艺术汇聚的胜地,在景观艺术设计方面的探索颇具参考与借鉴意义。

Jacob Javits Plaza

New York, NY, USA (1997)

In 1992, the Federal Government undertook the repair of the waterproofing for the underground garage beneath the Jacob Javits Plaza. Because the existing plaza would be demolished during the waterproofing construction, there arose an opportunity to revitalize the plaza. During the time that Richard Serra's "Tilted Arc" inhabited the plaza, this 14-foot-high sculpture was an obstruction both visually and physically to pedestrians. After the sculpture was removed, the plaza remained vacant and disconnected from its context. The intent of the plaza redesign was to create a useable, lively open space in the heart of the city. Full art and landscape architectural design services were required for this transformation to take place.

The new plaza is reconnected to its surrounding context and provides innumerable seating opportunities for people having lunch or just for watching others. Large planters which formerly existed at the northwest and southeast corners of the site have been removed, as well as the long, empty fountain which had occupied the only sunny portion of the site. By opening up the plaza, the connections between the plaza and the street are re-established, and the people who wish to sit can do so in either sun or shade.

The seating for the site is provided on twisting strands of New York City park benches. The double strands of back-to-back benches loop back and forth and allow for a variety of seating—intimate circles for groups and outside curves for those who wish to lunch alone. With their complex forms and bright green color, these benches energize the flat plane of the plaza in the same way that the French used the parterres embroideries which were punctuated by topiary forms and whose edges were defined by trees and buildings. The bright green color of the benches was selected because its reflectivity helps to enliven a plaza, which for the most part, is in shade.

At Jacob Javits Plaza, the benches swirl around the "topiary" or 6 foot tall grassy hemispheres that exude mist on hot days. Familiar lunchtime elements are provided such as blue enameled drinking fountains, orange wire-mesh trash cans, and Central Park lighting standards. While all of these elements are drawn from the Olmstedean tradition which maintains its hold in New York City, each element is tweaked slightly from its historic predecessor. These elements offer a critique of the art of landscape in New York City, where the ghost of Frederick Law Olmsted is too great a force for even New York to exorcise. The design itself offers a wry commentary on the fact that while New York remains a cultural mecca for most art forms, exploration in landscape architecture receives little support.

市政景观 Civic

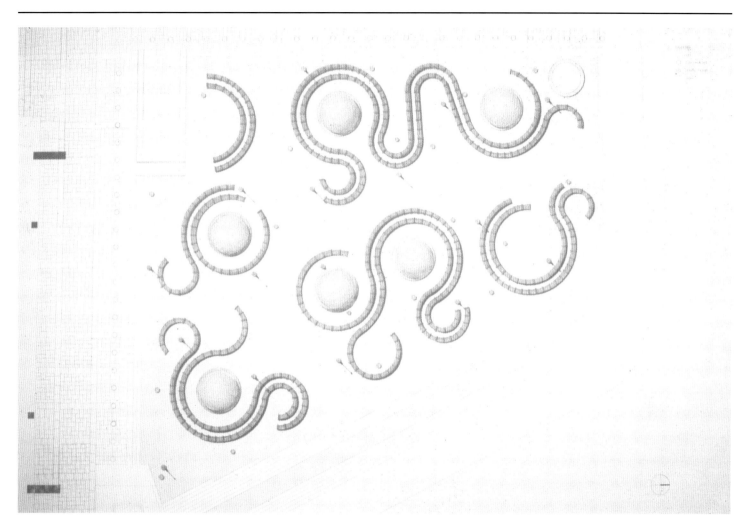

市政景观 Civic 158

市政景观　　　　　　　　　　　　　　　　　　　　Civic　　　　　　　　　　　　　　　　　　　160

林肯路购物中心

美国佛罗里达州迈阿密海滩，1987

林肯路购物中心于1960年设计，当时是一个步行购物中心。根据景观设计师莫里斯·拉皮德斯所提出的设计方案，出于步行需要，街道被封闭，禁止机动车通行，这在当时众多的美国购物中心中还是第一个。原先的购物中心长1372米，配有拉皮德斯结构的遮阴、喷泉和植被，以及商店、餐馆、酒吧和展览馆。不幸的是，20世纪70年代，曾经热闹、时髦的购物中心处于失修状态。20世纪90年代初期，在南海滩市复兴改造运动的推动下，沿街新建了艺术展览馆，这个购物中心的改造更新被重新提上了日程。

在设计方案中，在30.5米宽的街道两侧重新种植棕榈树，并移除那些已经死亡的棕榈树。如果没有这些"阻碍"，道路两侧只有低矮的两层建筑，那么，这里就缺乏封闭感，整条街道变成一个宽广的开放空间。随着时间的推移，步行道开裂，花草凋谢，路面碾压破损严重。

改造后的道路提供了更多的遮阴和开放空间。双排棕榈树具有视觉连续性。购物中心包括11个区块，通过地面铺装图案的变化，每一个区块都有自己的特征。地面铺装图案的设计灵感源自区块中原有的拉皮德斯结构。例如，"之"字形的地面铺装图案取材于"之"字形的拉皮德斯结构，并贯穿整个小区。拉皮德斯结构的历史构筑物被重新翻修。一系列喷泉被加以改造，并新增一处喷泉。一些新的物种被用来替换种植容器中原有的植物，并作为每一块区块的主题元素。

Lincoln Road Mall

Miami Beach, FL, USA (1987)

Lincoln Road Mall in South Miami Beach was designed in 1960 as a pedestrian mall. Conceived by Miami architect Morris Lapidus, the street was one of the first in America to be closed off to vehicles for this purpose. The original 4,500-foot-long mall incorporated Lapidus's fantastic shade structures, fountains, and plantings, flanked on both sides by shops, restaurants, bars, and galleries. Unfortunately, by the 1970s, the once-fashionable and upscale mall had fallen into disrepair; the revival of South Beach in the early 1990s brought new art galleries to the street and a renewed interest in renovating the mall.

Lapidus designed the original 100-foot-wide street to be planted on both sides by palms, which had long since died and been removed. Without the trees, the street was a wide open space that lacked a sense of enclosure by the low, two-story buildings along its edges. Over time, the sidewalks cracked, plantings changed, and the street became badly run down.

The new scheme provides more shade, contains the space, and creates a visual continuity by replanting native sabal palms in double rows on both sides of the street. Each block of the 11-block long mall is also given its own identity through a paving pattern that is generated by the Lapidus structures that inhabit the block. For example, the zigzag structure generated a zigzag paving that extends through the block. The historic Lapidus structures are being refurbished, existing fountains are being renovated and new ones added, and planters are replanted with new species that reinforce each block's theme.

| 市政景观 | Civic |

交易广场

英国曼彻斯特，1999

交易广场，位于英国曼彻斯特市中心。广场的周边是热闹非凡的零售和休闲娱乐区。为了迎接新千禧年的到来，广场进行了一系列开发建设。过去十多年以来，这一地区对整个城市的重生起到了催化作用。

广场设计的关键在于，广场向外一直延伸至建筑边缘，足以影响周边建筑和街道上的交通和行人出行。考虑到场地地形条件，水平方向的地形变化是设计重点之一。水平方向的地形变化有助于营造多元化的活动场地。广场作为周边建筑的背景，方便人们在此活动。

大量的坡道和台阶把两块高低不等的地面串联起来，运动和静止不断更替。车道作为尺度适宜的户外景观，既可以通行，又可以临时停放车辆。

广场的上方是这里一片面积最大的开放区域，主要零售活动都集中在这里。在这块平面场地中，嵌入式铁轨和彩色玻璃板从下方提供照明。铁轨标志着在曼彻斯特工业发展过程中铁路的重要性。成排的彩色照明将广场带入21世纪。在一年中的大部分时间里，曼彻斯特都笼罩在黑暗和多云之中。照明的有效使用充满神秘色彩且非常具有戏剧性。轨道上的活动座椅可以根据需要随时调节。

场地中较低矮的部分，光照充足，靠近喷泉，可以在此举行各种户外就餐活动。历史悠久的悬渠通过一条特殊的河流重新获得生命力。人工挖掘的一条小溪，用踏步石和水填充。溪边的弧形喷头中，喷出的水注入小溪之中。河桦树作为水景线的标识，给人一种舒缓惬意的感觉。

Exchange Square

Manchester, UK (1999)

Exchange Square is located in the heart of Manchester, UK. The square is at the heart of a vibrant retail and entertainment district which was developed for the turn of the millennium, over the last 10 years this district has catalysed the regeneration of the entire city centre.

Vital to the design of the square is that the plaza extends out to the building edges as its success, in part, is due to carefully "borrowing" the activity of the surrounding buildings and streets. Because of the existing topography, the sculpting of a plaza level change is the major design factor. The level change accomplishes three things. It creates places for a great variety of activities, it provides a setting for the surrounding buildings and it makes the square accessible to all.

Connecting the two levels is an exuberance of ramps and stairs that become objects of both movement and stasis. These ramps act as landscape-scale furniture, accommodating movement and informal seating.

The upper level is the largest open area of the site as well as where the majority of retail activity will take place. Inserted into this plane are flush-mounted rail tracks with inset colored glass panels lit from below. The tracks mark the historical importance of railroads in the industrial development of Manchester. The lines of colored light move the project into the 21st century. In a city that can be dark and overcast for much of the year, the effective use of light is a dramatic and wonderful addition to the public realm. Sliding along the tracks are moveable benches that allow seating to be rearranged as needed.

The lower level is the area of the site that will get the most sunshine and will accommodate outdoor dining with a close relationship to the fountain. The historic line of Hanging Ditch is brought to life through an abstracted river. An excavated "ditch" is filled with stepping stones and water. Arcing jets spray water along and over the stream. River Birch trees mark the line of the water feature giving a soft and more casual quality to this area.

Civic

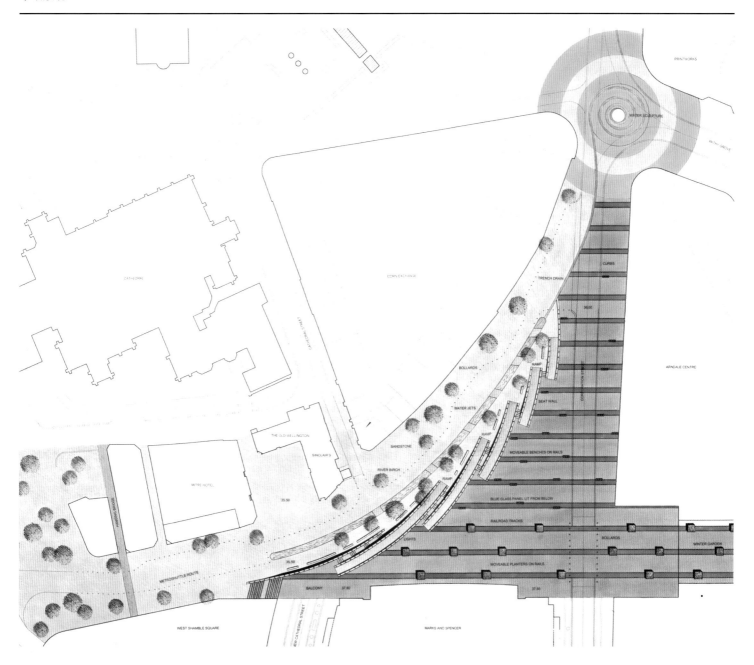

佛里斯顿村庄绿地

英国卡斯尔福德，2004

佛里斯顿村庄绿地的景观设计经过一系列公众咨询会议中的讨论研究而最终确定。这些讨论研究使设计团队对场地现状、当地居民的需求和期望有了比较深入的了解。结合公众咨询会议所阐明的场地功能与特征，设计团队提出了多个可供选择的设计方案。以这些设计方案为基础，设计团队可以继续与当地居民展开对话。

在佛里斯顿，马匹是日常生活中的重要交通工具之一。设计团队打造了一条专供马匹通行的道路，其与村庄东面和西面与马匹有关的基础设施串联在一起。这条小路的两侧有一系列矮墙，对马匹进行引导；同时，它们也是小路与北街的"天然屏障"。马匹通行道与绿地在中间处断开，这里有出自艺术家安东尼·葛姆雷之手的四组路桩。

游乐区位于一片平整的洼地之上。洼地四周的坡面呈台阶状，比较陡峭，可以作为座椅，或在此开展非正式的小型娱乐休闲活动。游乐设施位于场地的平坦地带，适合各种年龄的人群。两个区域通过短坡面相连。在场地西南角、靠近游乐区的区域中，一道土丘把绿地围封起来，在视觉和实体上形成一道屏障。

游乐区坡面的对面是堆石，其正面形态与游乐区背面坡面形成鲜明的对比。由琢石和巨砾堆积的螺旋状石堆逐渐变小，构成系列玻璃砖。玻璃砖在灯光的照射下构成绿地的视觉焦点，无论白天还是黑夜，都吸引着人们的注意力。除了堆石灯光之外，四根灯杆也可用于照明，避免夜晚发生潜在的危险，延长场地可使用的时间，特别是在冬季每天下午4~5点黄昏之时。绿地上新安装的照明设施保证这个新建社区一年四季充满活力。

应当地居民的要求，景观植物以高大的树木为主，而并非灌木。在生活安全和道路安全方面，当地居民希望视野不受任何阻挡。这些高大的树木多为当地特有的橡树（又名"欧洲栎"），按规整的行间距有序地排列。

Fryston Village Green

Castleford, UK (2004)

The design process for Fryston Village Green was based on a series of public consultation meetings that were vital in helping the design team understand the context of the site and the needs and aspirations of the current residents. These public meetings dealt with the site as well as the residents' preferences in terms of design vocabularies. The results of this process were combined with the functional and aspirational requirements for the site and led the design team to generate a series of design options. These design options continued to provide a basis for dialogue with the residents.

As horses are an integral part of daily life in Fryston, the design features a horse trail that connects the equine facilities on the east and west of the village. This trail is bordered on both sides with a low wall to guide the horses, provide informal seating, and create a barrier from North Street. A central break between the trail and the Green is marked by four bollards designed by artist Antony Gormley.

The play area of the Village Green is formed by a flat area and a depression in the ground plane. The side of the depression forms a play slope edged with deep steps which act as seating. Play equipment designed to appeal to a range of ages is located on the level area. The two areas are connected by slides down the short slope. Adjacent to the play area in the southwest corner of the site is a mound which encloses the Green and creates visual and physical containment. The sloped sides also provide more opportunities for informal play and seating.

The inverse of the play slope is the rock pile Cairn—this positive form contrasts with the play area's negative slope. The Cairn's spiral mound of cut stone and boulders thins out into a series of glass bricks. These bricks are lit to provide a focal point for the Green that draws attention both day and night. In addition to the lighting of the Cairn, four light poles brighten areas of potential night time hazard and extend the hours during which the site can be used, especially in the winter months when dusk occurs at 4 or 5 o'clock. New lighting on the Village Green ensures this new communal space will be an active facility throughout the year.

Landscape plantings are used selectively focusing on large, high-quality tree specimens rather than banks of shrub planting. This decision was driven by local concerns in which residents preferred unobstructed views for security and road safety. The trees are native oaks (Quercus robur) to continue with a clear planting scheme.

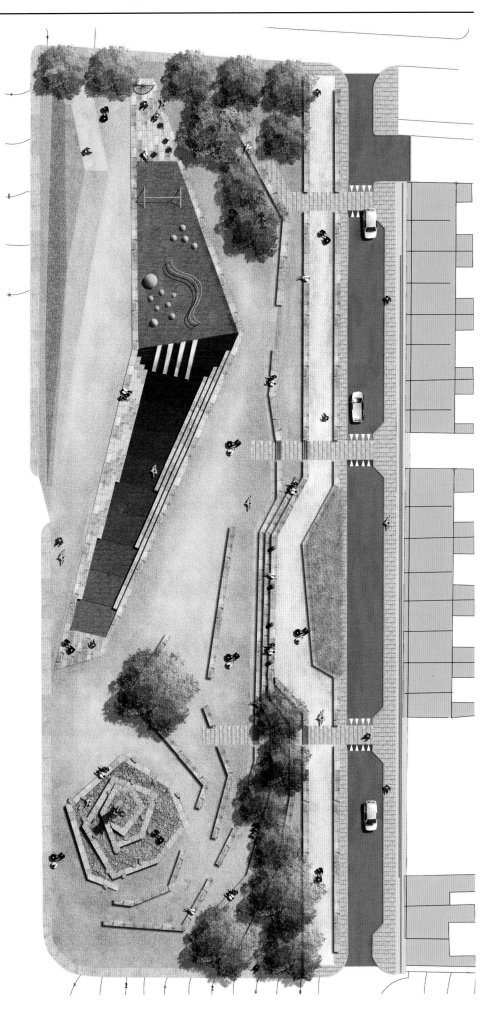

市政景观 Civic 172

市政景观　　　　　　　　　　　　　　　Civic

梅萨艺术中心

美国亚利桑那州梅萨，2005

场地设计的中心思想是打造一条宽敞的步行道，横贯整个场地，同时为各种大型和小型团体聚会提供场所，供人们安静地休息放松和赏景。设计的主题是"林荫步行道"，各种树冠和建筑的顶盖形成了相互交叠的阴影，营造了迷人的环境氛围。沿着步行道，由树木构成的曲线形线条，前后不断变幻，在地面上形成各种不同的阴影，形态和大小变幻莫测。除了植物所形成的阴影之外，各种彩色玻璃所形成的冠状结构以及被抬高的玻璃屏幕也会在地面上投下一系列彩色阴影。

彩色玻璃具有双重作用，从上面向人行道喷雾，用来降温。每日午后，阳光穿透半透明的彩色玻璃幕墙，使仙人掌类植物以及其他个性鲜明、质地不同的植物投射形成各种不同的阴影。

与"林荫步行道"主题相平行的是通向西南角的水景：一条小渠，里面填充砾石，沿着"林荫步行道"西侧，横贯整个场地。强大的水流时不时地从北向南冲刷着河床，提醒人们这块场地具有山洪特征。冲刷间隙，水蒸气从潮湿的砾石上向外蒸发，营造一种凉爽和潮湿的效果。

与"林荫步行道"相关的另一个主题是"宴会桌"。用于捕捉阳光的彩色玻璃投下各种不同的阴影，雕塑般的桌椅也形成一系列彩色阴影。这些阴影中的"宴会桌"供人们举行各种庆祝活动，梅萨艺术中心充满诗一般的意境。

Mesa Arts Center

Mesa, AZ, USA (2005)

The central idea of the site design is to provide a grand promenade into and through the complex while providing opportunities for both large and small group gatherings, as well as places for quiet relaxation and enjoyment. The theme is a "Shadow Walk": a place where the rich interplay of overlapping shadows, trees and architectural canopies create a cool and inviting environment. Long, curving lines of trees, shift back and forth as one walks along the promenade, throwing different shadow forms on the ground and creating different qualities and quantities of shadow. In addition to shadows thrown by vegetation, a series of colored glass canopies and raised glass screens will cast colored shadows on the ground.

These colored glass structures serve a double purpose in that they are also the structure for cooling mist jets that spray down to the walkway from above. Translucent colored glass screen walls, back-lit by the afternoon sun, will hold the shadows of cacti and other distinctly textured plants in silhouette.

Paralleling the Shadow Walk theme is a water story appropriate to the southwest: a boulder-filled arroyo runs along the western side of the Shadow Walk for its entire length. From time to time, a strong pulse of water will rush through the riverbed from north to south, recalling the flash floods characteristic of the region. A welcome side benefit of this exciting event will be the cooling and humidifying effect of the water evaporating from the wet boulders in the interval between episodes.

Another motif running through the Shadow Walk is that of the banquet table. The colored glass forms used to catch light will cast shade, and color shadows will also take the form of sculptural and symbolic tables and chairs. These forms are abstracted to create a poetic statement about people coming together in celebration, a perfect family-oriented image for the Mesa Arts and Entertainment Center and for the heart of the community.

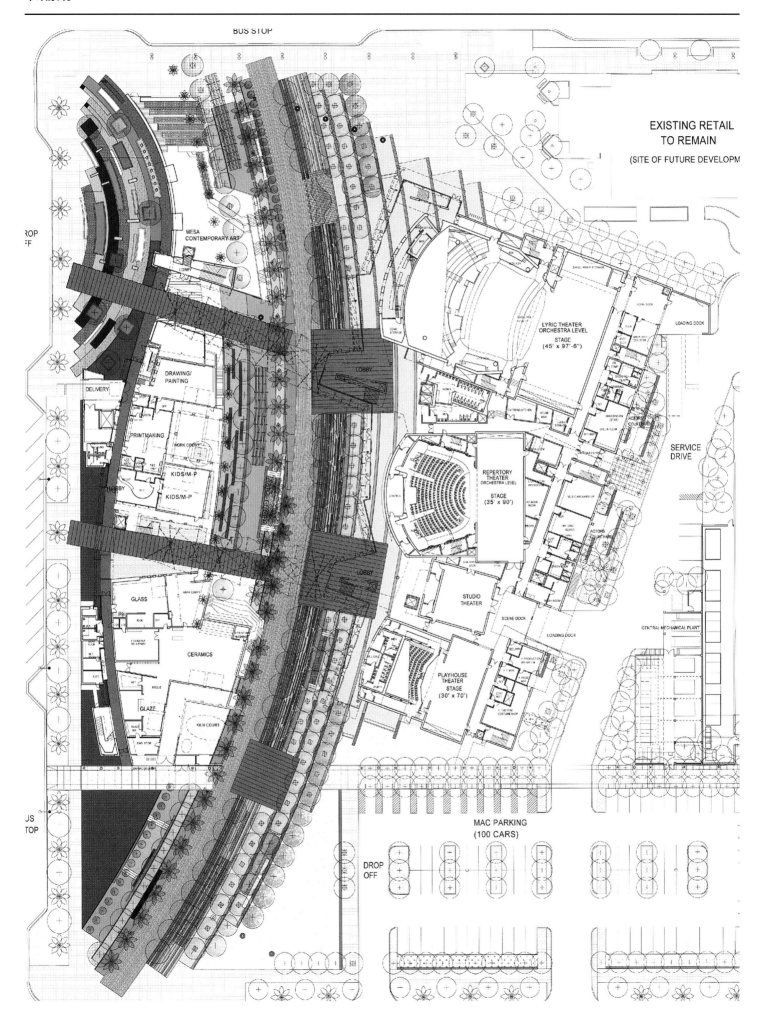

市政景观 Civic 176

市政景观 Civic

市政景观　　　　　　　　　　　　　　　　　　Civic

大运河广场

爱尔兰都柏林，2007

大运河广场的设计灵感源自场地景观现状和丹尼尔·里伯斯金工作室所设计的大剧院，这家剧院于 2010 年建成。这个投资 800 万英镑的大运河广场取代了原来的小型铺装广场，其下方是一个地下停车场。

在都柏林港口开发区，大运河广场是最主要的公共开放空间。它位于都柏林大运河之畔，构成这个新开发区的视觉焦点。广场的一端面向宽广的大运河，另一端有新建的里伯斯金著名的大剧院和娱乐休闲建筑，北面有新建的宾馆，南面是办公建筑。在这块现代建筑占主导地位的场地之中，景观设计师打造了一块公共空间，为都柏林的开放空间增添色彩并注入活力。考虑到其著名的文化特征，景观设计师"铺设"了一块中央红色地毯，其从剧院向外一直延伸至运河，反之亦然。一块绿色地毯把新建的宾馆和办公建筑串联在一起。

硬质铺装景观主要由碎玻璃构成，形成一系列十字交叉的"希望之路"，贯穿整个广场，把各个重要节点连接起来。原来的广场太小了。景观设计师试图扩大广场规模，把硬质铺装景观一直延伸至建筑边缘。原先广场中的花岗岩地面铺装在新设计方案中得到重新利用，以修建穿越广场且通往各个方向的小路。广场提供了充足的空间，可在此举办各种公共活动，比如节日庆典和表演活动。

红色地毯，从剧院台阶开始，穿过广场，把剧院的魔力散布于公共空间之中，向下一直延伸至水边。地毯由鲜红色树脂玻璃构成，白天反射和捕捉阳光。嵌入地毯中的红色拾取棒，在夜晚散发出戏剧般的光芒。

绿色地毯让人感到平和，并在高度各异的种植容器边缘提供充足的座位。种植容器，是由绿色地毯构成的拉伸多边形，湿地植物遍布其中，提醒人们这块场地曾经是一片湿地。优雅的草坪可供人们散步，欣赏这片壮丽的景观。广场外面为水景，水从随意堆放的绿色大理石中缓缓冒出。狭窄的小路，十字交叉，进一步对广场进行划分，游客可以从任何方向穿过广场。同时，广场也可作为市场或者集市，在此举办各种大型活动。

广场上有两个三角形的小亭子，它们蜿蜒地通往地下停车场，作为通道，为周围空间增加了通风。这一路上，风景变幻莫测。小亭子的外檐由蓝色墙面构成，其中镶嵌几盏蓝色 LED 节能灯。小亭子被水景和山脉所环绕，夜晚，在蓝色 LED 节能灯的映衬下，水面熠熠生辉，山脉愈发神秘。

大运河广场作为都柏林的标志性景观，就像一块城市磁铁，吸引大批国际化的公司入驻此地，同时，当地居民也喜欢来到这里纳凉聚会、娱乐休闲，这大大促进了当地经济发展和邻里交流。

Grand Canal Square

Dublin, Ireland (2007)

Inspiration for Grand Canal Square came from the existing landscape and the theatre by Studio Daniel Libeskind completed in 2010. The € 8 million project has replaced a smaller paved space built over an underground car park.

Grand Canal Square is the major public open space in the Dublin Docklands Development area. It is located on Grand Canal in Dublin and forms the focal point of this development. In a setting dominated by contemporary architectural expression, we have created a public space that will offer color and dynamism to Dublin's open spaces. Due to its cultural celebrity setting, we have developed a scheme with a central red carpet that leads from the theatre out onto the canal and vice versa. A green carpet connects the new hotel to the office development.

The hardscape consists of a cracked-glass scheme—a series of criss-cross "paths of desire" stretching across the length of the square connecting various points of interest for pedestrians. The original square was much smaller. MSP wanted to enlarge the square by extending the hardscape up to the buildings' edges. Granite paving from the previous square, laid out just two years previously, was been recycled in the new design to create paths across the square in every direction while still allowing for the space to host major public events such as festivals and performances.

Extending out from the steps of the theatre, the red carpet rolls into the square, spilling the magic of the theatre into the public space and down to the water's edge. The carpet is made from bright red resin-glass pavings that reflect and capture light during the day. The red pick-up sticks imbedded into the carpet provide dramatic light at night.

The green carpet has a calmer expression and offers ample seating on the edges of planters of various heights. The planters, extruded polygons of the green carpet, are planted with marsh vegetation as a reminder of the historic wetland area of this site and some offer immaculate lawns for lingering and enjoying the spectacular setting. Pushing out of the plaza is a water feature of randomly stacked green marble that is overflowing with bubbling water. The square is further criss-crossed by narrow paths that allow for movement across the square in any possible direction while still allowing big activities such as markets or fairs. The new square will be an urban magnet with 24-hour activity and is an accurate interpretation of Dublin's energy.

The three triangular objects are the two pavilions and the water feature. The two pavilions allow access to the stairs and elevator leading to the underground parking as well as add ventilation. The pavilions are stainless steel mesh with blue walls and blue LED lights. The third triangle, a cascading fountain, evokes a boulder being pushed from the ground with different layers representing a "brook in spring". White LED lights illuminate the water at night.

Grand Canal Square has helped to create a presence that gave the area's development community confidence to move forward. The space created a strong address and presence and is now the address for several international company headquarters. It helped to keep this neighborhood thriving during an economic downturn by creating a desirable area in which to work and live.

市政景观 Civic

市政景观　　　　　　　　　　　　　　　　Civic　　　　　　　　　　　　　　　　185

市政景观 Civic

市政景观 Civic

市政景观 Civic

| 市政景观 | Civic | 190 |

贝鲁特滨水公园设计竞赛

黎巴嫩贝鲁特，2011

黎巴嫩贝鲁特开发重建公司邀请玛莎·施瓦茨及合伙人设计事务所参与贝鲁特滨水公园设计竞赛。设计对象包括贝鲁特滨水公园和滨海长廊。这条长廊把原有的长廊体系延伸至城市中心。贝鲁特滨海新区属于贝鲁特中心城区开发的第二阶段。这块填海所形成的土地，面积73公顷，周围是阶梯式海防体系，能够经受百年不遇的暴风雨。开发建设的目标是在既定的指导原则下，将这块土地打造成一流的多功能新区，其中包括大面积的绿地和特征鲜明的建筑，并成为地区可持续性发展的典范。

这个滨海新区，北面，面向地中海，视野开阔；东北面，横跨圣乔治湾的山脉，有良好的视野。滨水公园被建在"战时垃圾场"（又称"诺曼底垃圾填埋场"）上。这个公园作为中心城区一个重要的娱乐休闲场所，在其内部和附近有各种娱乐休闲和文化设施。滨海长廊穿过整个滨海新区，构成一条连续的阶梯式公共廊道，作为全市范围内的活动场所：整个滨海长廊体系环绕贝鲁特半岛。

景观设计师通过对场地的深入了解，试图彰显出贝鲁特滨水公园的本质。这个项目的核心问题是对土壤层次的探索。也就是说，公园建设所使用的土壤讲述了有关贝鲁特的故事。通过土壤分层，这个独特的公共空间设计，连接着贝鲁特的历史和未来。在城市结构被改造并重新组合之后，一座新城脱颖而出。

公园深入地下，通过土笼，揭示出贝鲁特的过去。其上展示着贝鲁特的未来。公园的空间朝向也充分利用场地的地形条件，打造一个高出海平面的平台，可俯瞰贝鲁特城和群山。设计方案提高了贝鲁特的城市可持续发展能力。在这里，人们之间以及人与自然之间可自由地交流互动。同时，各种动植物在此"休养生息"。这里堪称"生物多样性环境保护"的典范之地。

Beirut Waterfront Park Competition

Beirut, Lebanon (2011)

The Lebanese Company for the Development and Reconstruction of Beirut Central District, invited MSP for a limited concept design competition to engage in the development of a park on Beirut's New Waterfront District as well as the Corniche Promenade; an extension of the existing corniche system into the city center. The New Waterfront District is the second phase of Beirut City Centre development. The 73-ha area of reclaimed land enclosed within a terraced sea defense system designed to withstand centennial storms, destined to be a prime, multiuse district with extensive green areas, distinctive architecture and developed with mandated guidelines as a model project of sustainable urban development in the region. The Waterfront District commands fine views of the Mediterranean to the North, and the mountains across St George's Bay to the Northeast.

The Waterfront Park, built on the former wartime dumpsite, known as the Normandy landfill. The new park is one of the main public attractions in the City Center, with planned recreation, leisure and cultural facilities within it and in the vicinity. The Corniche Promenade forms a continuous, terraced public promenade across the entire New Waterfront District, providing a city-wide attraction: the culmination and destination of the entire corniche system that fringes the Beirut peninsula.

MSP sought to express the essence of the Beirut Waterfront Park by understanding the very land it was built on. The exploration of the land strata is at the heart of MSP's thinking for this project and hence, the soil from which the park was created tells the story of Beirut. The design of this unique public space has been expressed in a way that allows the stratification of the earth to convey the history and the future of Beirut. The fabric of the past and existing city creates a new city.

The park cuts through the ground, exposing layers of Beirut's past which are captured in the gabions; above which, sits the future. The spatial orientation of the new park also harnesses the topography which creates a platform that rises above the sea and makes a vantage point overlooking the city and mountains of Beirut. The design also encompasses and enhances Beirut in a sustainable way that allows social interaction and spaces for people and the natural environment—ecological habitats are created in the urban plaza that encourage the creation of wildlife and bio-diverse environments.

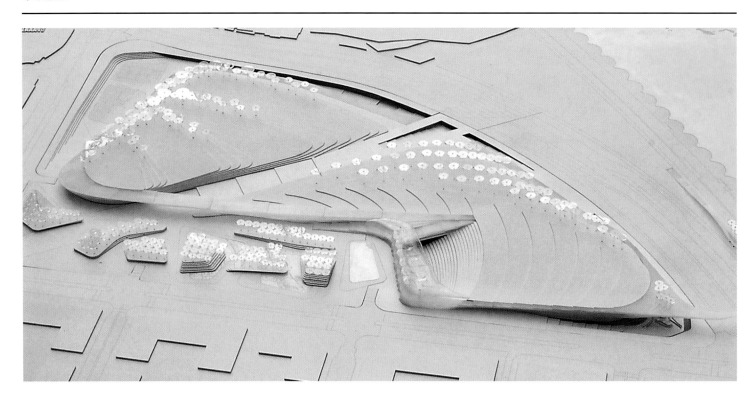

市政景观 / Civic

索沃广场

阿联酋阿布扎比，2012

在阿布扎比索沃岛大型总体规划中，索沃广场是首先开发建设的项目，是整个索沃岛的开发中心，也是城市商业中心的绿色缓冲地带。

广场的设计灵感和理念来自阿拉伯半岛与生俱来的自然和文化特性：沙丘、传统灌溉系统（法拉）、绿洲、贝都因纺织品、修剪规整的绿篱、法国巴洛克风格的别墅花园。在这个广场的设计方案中，这些元素得到了很好的体现。令人倍感舒适的可持续性微气候，动态的万花筒式植物种植布局，带有多种图案的地面铺装，从周围高楼大厦上所俯瞰的壮观景色，这一切为游客营造了良好的娱乐休闲空间。

在整体结构上，广场采用带有植被的大型结构性沙丘，构成景色优美的户外空间，保护行人，防止夏马风（吹向波斯湾的强西北风）的吹袭，在高楼大厦林立的环境之中营造视野开阔空间。装饰性的地面铺装，就像传统地毯，把这些沙丘连接在一起，在广场上波动起伏。为了有效降低温度，使游客感到舒适惬意，一些长长的石凳围绕着山丘，长凳边有水景，这些景观都是可感触的。长凳表面刻有造型美丽的沟槽，就像一片片动态的涟漪。为了充分利用水源，减少蒸发，水通过狭窄的法拉（中东地区随处可见的古代灌溉系统，类似于水渠）流动。夜晚，底部安装照明灯的长凳散发出光芒，映衬着沙丘的轮廓，长凳表面也熠熠生辉。

广场设计运用可持续性设计思维，最终，设计获得了 LEED 金奖。带有植被的大型结构性土丘所营造的绿地空间几乎是种植容器所营造的绿地空间的 1.45 倍。得益于土丘周边的垂直绿化，广场周边一直保持湿润的微气候，水资源消耗也大大减少。

Sowwah Square

Abu Dhabi, UAE (2012)

Sowwah Square is a significant urban space located on the Al Maryah Island, Abu Dhabi and provides a green retreat at the centre of the new commercial hub.

The inspiration for the square was derived from the nature and culture inherent to the Arabian Peninsula: dunes, traditional irrigation systems (Falaj), oasis, Bedouin textiles and the popular use of formal clipped hedges in United Arab Emirates, drawing connections with the French baroque château gardens. This merging of ideas is represented in a contemporary responsive design creating a sustainable, cool and protected micro climate for users and a dynamic kaleidoscope of planting and patterned paving on the ground and viewed from the surrounding towers.

The structure of the square uses large constructed vegetated mounds that orchestrate outdoor rooms to shelter pedestrians from the Shamal, a strong north-westerly wind blowing over the Persian Gulf and to provide intimate spaces amongst the towering buildings. Linking the mounds together, the decorative pattern like that of a traditional rug, weaves through the square. To soothe people from the heat, water features are incorporated into long stone benches that wrap the mounds, providing a playful and tactile experience. The surface texture is finished with ornate grooves creating a dynamic rippling effect. In order to maximise this limited resource, and reduce evaporation, the water is contained in narrow Falaj-like channels as used in ancient irrigation system found throughout the Middle East. At night, the benches come to life with integrated lighting at the base that silhouettes the mounds and highlights the polished surfaces.

Innovative sustainable design has been instrumental in the project which has been awarded a LEED Gold certification. The steep angulated mounds contribute 1.45 times more green space than level planters and water consumption is reduced due to the vertical planting maximizing 100% irrigation moisture.

市政景观　　　　　　　　　　　　　　　　　Civic　　　　　　　　　　　　　　　　　195

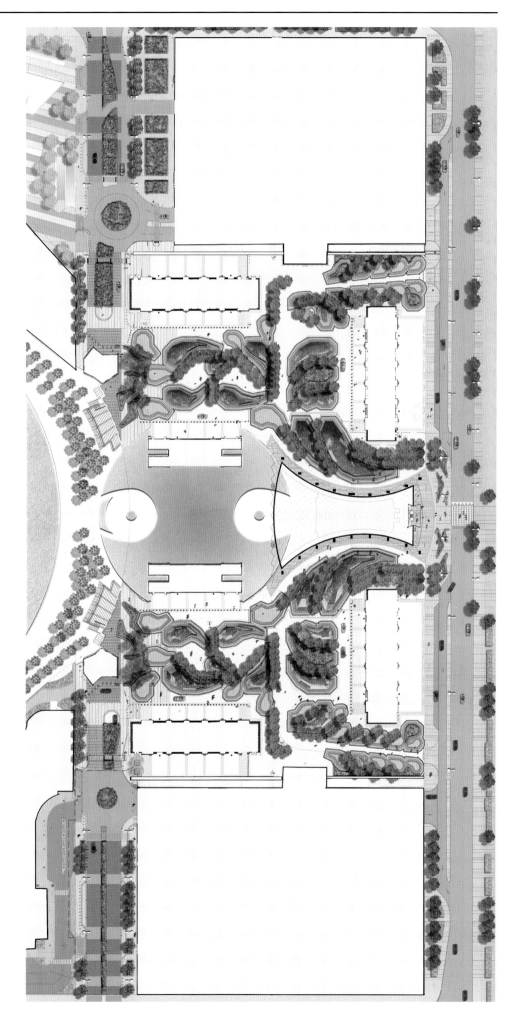

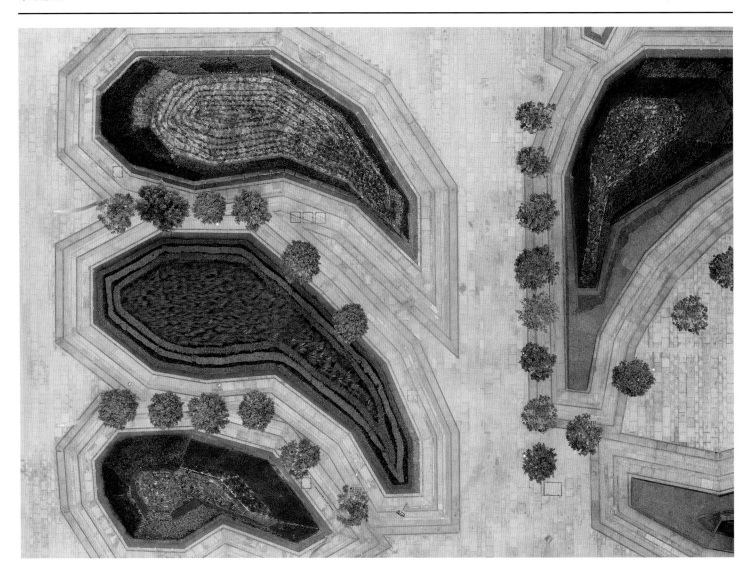

市政景观 Civic 198

| 市政景观 | Civic | 200 |

海军码头公园设计竞赛

美国伊利诺伊州芝加哥，2012

长期以来，芝加哥的城市形态沿着密歇根湖逐渐形成，主要围绕三大目标：保护密歇根湖的生态环境，促进经济增长，为市民创建可供生存发展和娱乐休闲的城市空间。原有的小型码头经过扩展改造，成为一个世界级的 21 世纪大型滨水码头。这个码头由五个部分组成：

水面跨越——重新设计入口公园，将其打造成动态城市景观，这里有新建的喷泉、颇具吸引力的景点以及位于南端的结构性湿地。

边缘强化——南码头新建一座长廊，增加一个浮动码头，使这个海军码头重新面向太阳、密歇根湖和芝加哥的天际线，从而带来神奇的变化。

上上下下——码头公园围绕着标志性景观"舵轮"雕塑和新建的莎士比亚剧院，起伏变化，重新焕发生机。

透镜观赏——水晶园经过设计改造，成为"人间仙境"。游乐区与码头公园和儿童博物馆相连。

冒险尝试——东端公园设有蒸汽浴池、冷水浴池以及巨石和沙滩，密歇根湖的天然特质与野性被充分融入公园之中。

重新设计的大型滨水码头使海军码头得以扩展，由线性通道变成多维通道。芝加哥市由此向外延伸，直抵湖岸线，由湖岸线向下与水面相接，再通过水面指向天际线。沿伊利诺伊大街，新建的长廊和灯塔向西延伸，再加上新设立的充气豆荚系统（码头豆荚），这座海军码头与芝加哥的连接得以加强，并且彼此之间保持一定的距离。海军码头公园力求成为富有创新精神且带有城市特色的滨水景观，并成为一流码头公园中的典范。

Navy Pier Competition

Chicago, IL, USA (2012)

Chicago's coastline has long been shaped by three ambitions: protecting Lake Michigan, promoting economic growth, and creating spectacular civic spaces for the pleasure and enrichment of its people. These same ambitions have inspired our team to expand Pierscape to PIERESCAPE: a world-class, 21st century waterfront experience with five uniquely imaginative episodes:

Crossing the Water—Gateway Park is redesigned as an active urban landscape with new fountains, attractions, and constructed wetlands at its southern end.

Taking it over the Edge—South Dock is dramatically transformed by the creation of a new Porch and the addition of Floating Piers that reorient Navy Pier to the sun, Lake Michigan, and Chicago's skyline.

Getting up to Get Down—Pier Park is reignited with a new undulating landscape surrounding the iconic Ferris Wheel and a new Shakespeare Theater.

Going through the Looking Glass—Crystal Garden is reimagined as a "wonderland", a play-scape connecting Pier Park and the Children's Museum.

Taking the Plunge—East-End Park dissolves into the wild majesty of Lake Michigan among steam baths, cold pools, boulders and beaches.

The PIERESCAPE expands Navy Pier from a linear to a multidimensional journey that connects city out to shoreline, shoreline down to water, and water up to sky. By extending the new Porch and Light Towers to the west along Illinois Avenue and creating a new PierPod aerial gondola system, the Pier becomes more connected to the city, yet also remains a far out retreat. We enjoyed working with the people of Chicago on realizing this vision for the lakefront that is expansive, innovative, sustainable, achievable and civic—all aspects of a world-class icon that is quintessentially Chicago.

市政景观　　Civic　　201

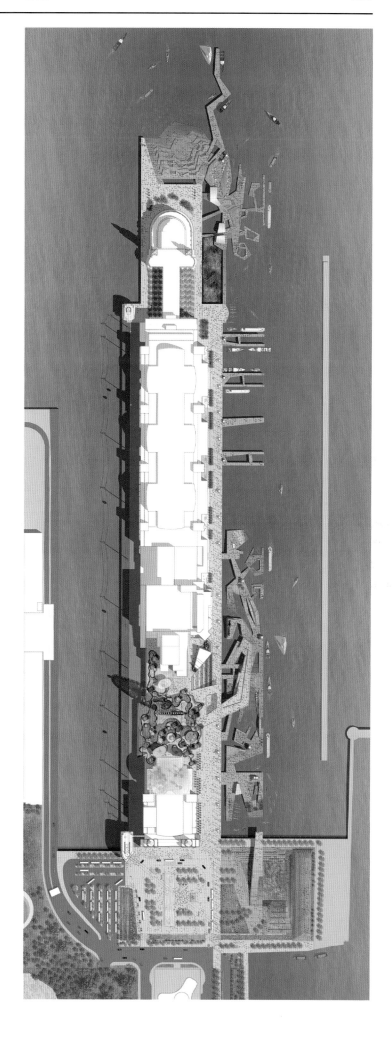

市政景观　　　Civic

| 市政景观 | Civic |

共和广场

法国巴黎，2013

共和广场曾经是令人震撼且充满活力的城市广场。现在，它已经被城市交通需求所击溃了。在这块场地上，每天有 114 000 名地铁乘客过往，有成排的旅游大巴车、出租车、自行车、小汽车以及公共汽车。这个广场变得拥堵、破碎、不安全，失去了地域特色和城市重要性。

广场设计力求使广场与周边基础设施相关联，提高容纳流动人口的能力，为非流动人员和设施留出一定的空间。场地中原来不相关的部分被统一起来，围绕中央核心区，重新设定机动车运行路线，把广场改造成一个大型功能性地表空间，满足各种活动和交通运行需求。每一项设计都力求使广场的功能最大化。嵌入式基础设施和可变空间使广场在各种活动之间实现无缝转变：从城市林荫大道，到周末市场、夜晚市场、圣诞市场；从摇滚音乐会，到电影节、盛夏节日和冬天的滑雪节。当地居民、上下班人员以及游客，日复一日、年复一年地重新回到广场上，寻找新的空间体验或重温往日的难忘经历。

这是一个可持续发展的城市空间。持续地适应不断变化的城市人口结构，并满足各种活动需要，把不断演化的交通运行模式整合在一起，为人们提供各种神奇精彩的空间体验。

Place de la République

France, Paris (2013)

Once a vibrant and active urban plaza, the Place de la République had been crushed by the demands of modern mobility. Each day the site hosts over 114,000 subway commuters and a complicated array of tourist buses, parked taxis, cyclists, car traffic, municipal bus routes, which have reduced it to a congested, fractured, and unsafe transitional space that has lost its local character and urban significance.

Our team's proposal restores the civic relevance to the plaza by improving its ability to sustain a moving population, while providing a reason to linger to a non-moving one. By unifying the disparate parts of the site and redirecting vehicular traffic away and around the central core, we have reclaimed a large functional surface that can support a range of activities and urban programs. Embedded infrastructure and variable spatial arrangements allow the plaza to seamlessly transition from urban boulevard to weekend market to evening market to Christmas market, from rock concert to film festival to summer festival to winter ice-skating rink. Residents, commuters and tourists will be able to return day after day, year after year, looking for new ways to experience the space and to relive old ones.

This is a sustainable city space. A space that continues to support the needs of an ever-changing urban population, that is endlessly adaptable, that integrates evolving urban transportation demands, and that continues to provide an open-ended stage for its visitors.

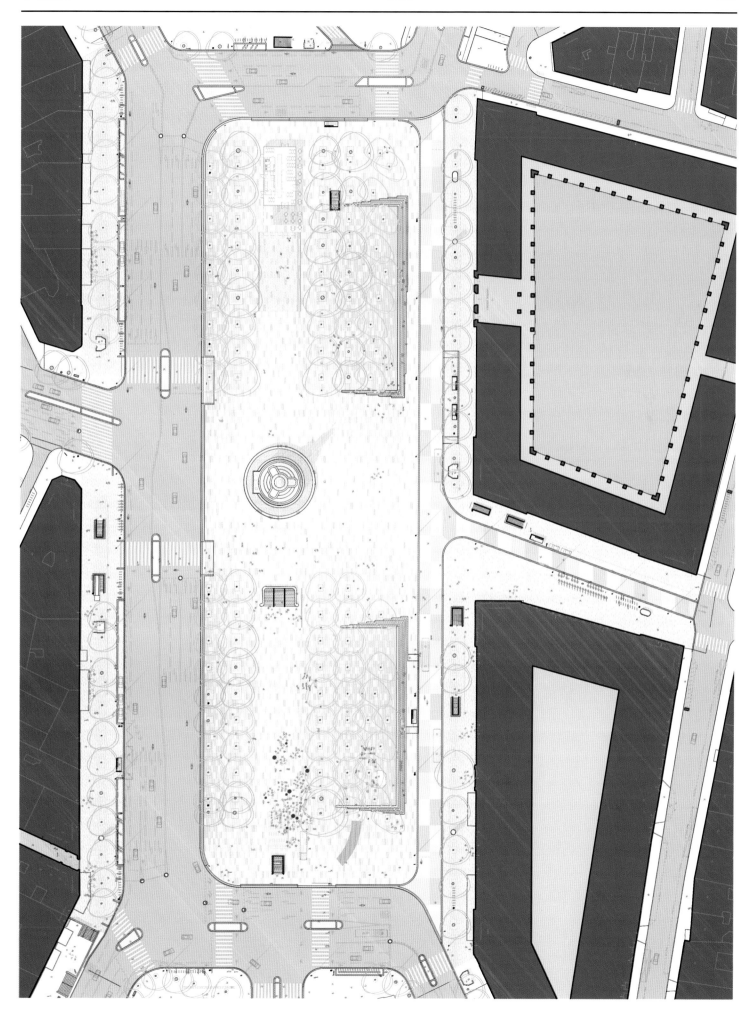

| 市政景观 | Civic |

北钓鱼台开发项目

中国北京，2014

北京长实东方置业有限公司委托玛莎·施瓦茨及合伙人设计事务所对北钓鱼台开发项目进行景观设计。项目位于北京以北50千米的雁栖湖畔。

项目包括一座温泉宾馆，由安藤忠雄设计，周边还有两个居住区。客户是唯一有资格承担此项目的私人开发商。雁栖湖，是2014年APEC会议会址。

这个项目在总体上属于沉浸式景观，把各种建筑元素整合在统一的环境之中，也就是把不同风格的建筑联系起来，形成一个协调的整体。除了总体景观规划之外，几个小区域的细部设计也是重中之重，包括范崎路（项目街道所在地）、南居住区出入口、北居住区出入口和岸边花园。

范崎路就像一条长340米的绘画透视模型，展现了中国风景画的多元和多变。从路边向里是一系列突出的石块，就好像一座"山"在向上攀登，对项目起到支撑保护作用。这些由石块组成的箱体作为种植容器，其中种植各种乡村树种、灌木和竹子。这些箱体就像一块块巨石，构成地质展示的"底座"。水流从中部宾馆入口处流出，沿着山坡向南、向北，漫过石块"底座"呈瀑布状倾泻，进入项目中心区域，象征着活力与财富。

North Diaoyutai Development

Beijing, China (2014)

Martha Schwartz Partners were commissioned by Beijing Changshi Oriental Land to produce a landscape vision for the North Diaoyutai Development at Yanqi (Swan) Lake, 50 km north of Beijing, China.

The project comprises a spa hotel designed by Tadao Ando, surrounded by two residential communities and a streetscape along Fanqi Road. The client is the only private developer to be granted a development licence at Yanqi Lake which was the site of the 2014 APEC meetings.

Our vision was for an immersive landscape framework, one that wraps the built components of the project within a unifying environment—tying together the varied architectural styles into one coherent whole. In addition to the overall landscape vision MSP was responsible for the detailed design of a number of areas within the project, these included, Fanqi Road (the street address of the project), the Southern and Northern Residential entries and the Shore Garden.

Fanqi Road is a 340-metre-long diorama—a contemporary interpretation of a Chinese landscape painting, illustrating the many and varied facets of the Chinese landscape. The landscape rises away from the road in a series of extruded stone boxes, a "mountain" stepping up and protecting the North Diaoyutai project. These boxes form planters for native tree, shrub and bamboo planting. The boxes also form "plinths" for the exhibition of Chinese geology in the form of massive rocks. Water runs from the central hotel entrance downhill to the south and north, cascading down the stone plinths and running into the development as a symbol of vitality and wealth.

市政景观　　　　Civic　　　　213

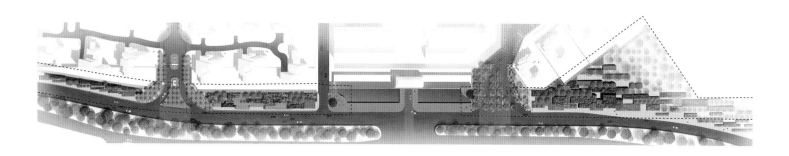

市政景观　　Civic　　215

| 市政景观 | Civic | 216 |

莫斯科儿童休闲街区

俄罗斯莫斯科，2015

玛莎·施瓦茨及合伙人设计事务所接受了斯特列尔卡公司公司的委托，对位于莫斯科中心城区的街道景观进行重新设计。项目范围覆盖了6.6公顷的街区，从东部的卢比扬卡广场，向西部和西北部延伸，著名的儿童中央商店和萨伏伊酒店都划归其中。目前，这些街区还不够人性化，道路昏暗且乏善可陈，稠密的地下服务网络和公共设施使得城市植被的效用大幅降低。在一个超过1200万人口的城市中，车辆交通不可避免地成为这个地区首要出行方式。

项目负责人深知，这个星球上最大国家的首都进行自我改造的重要性。出于这种认知，将这个项目的宗旨设定为将莫斯科改造成这样一座城市：出行舒适，孩子们可以尽情地互动嬉戏，街道功能满足不同需求的使用者，植被繁茂且空间共享，最大程度利用城市空间，减少机动车运行噪声，以及为步行者创建安全、舒适的环境。

为了这个目的，MSP设计了一个能使这些元素连贯统一、清晰易辨、和谐共生的体系。项目自始至终使用的是质地坚实，色调极简的材料。地面铺设颜色对比强烈的黑白花岗岩，并在地面搭配建设坚固的混凝土种植容器，黑色纹路的木质设施以及结实的花岗岩护桩。轻柔高雅的灯光辅助以强光来为街道设施照明。Pushechnaya大街和Rozhdestvenka大街被改造成共享空间，优先供给行人过往。植物被种植在大型的容器中，这克服了地下服务设施对植被影响的问题，为植物生长创建了最优环境，同时也缓和了俄罗斯冬季的寒冷。在共享区域里，这些种植容器就像一座座雕塑，周围有座椅和混凝土台阶，十分适合孩子们玩闹嬉戏。其他户外设施分散于各个区域，为居民们休闲玩乐创造了亲密的空间。

景观设计力求打造健康宜居、安全可靠、令人愉悦的街区环境，以满足21世纪城市发展的需求。

Moscow Children's Route

Moscow, Russia (2015)

MSP was appointed by Strelka KB to redesign the Streetscape of an area of central Moscow. The site covers 6.6 hectares of streets from Lubyanka Square in the east, spreading west and north-west, enveloping the famous Central Children's Store and the Savoy Hotel. These streets at present are not pedestrian friendly, dark and dreary, with a dense subterranean network of services and utilities negating the relief afforded by urban planting. In a city of over 12 million people, vehicular traffic has inevitably come to dominate the area.

As the capital of the largest country on the planet, the client understood it was important for the city to transform itself. From this conviction the competition brief evolved to transform Moscow into a city in which it is comfortable to walk, where children are encouraged to play and interact in the environment. The streetscape should be designed in a way to accommodate multiple uses, introducing planting and shared surfaces, expanding the possibilities available and calming vehicular movement, elevating the pedestrian in a safe and enjoyable way.

To this end MSP has designed a scheme which brings these elements together in a coherent, unified and legible manner. A strong and minimal palette of materials is used throughout. Granite paving in contrasting black and white for pavements and shared surfaces is paired with robust concrete planters, blackened timber furniture elements and sturdy granite bollards. Simple, elegant street lighting is supplemented with accent lighting in street furniture. The entire length of Pushechnaya Street and Rozhdestvenka Street are transformed into shared surfaces where priority is given to the pedestrian. Planting is accommodated in large planters overcoming the problem of underground services. These have been designed to provide the optimal growing conditions, and mitigation measures for the harsh Russian winter. In the shared surface areas these planters become sculptural elements, surrounded by seating and concrete stepping blocks suitable for children's play. Other furniture elements are scattered along these areas creating intimate programmes for rest or play.

MSP has endeavoured to create a flexible, safe and enjoyable environment, one that meets the needs of a 21st century city.

市政景观 Civic

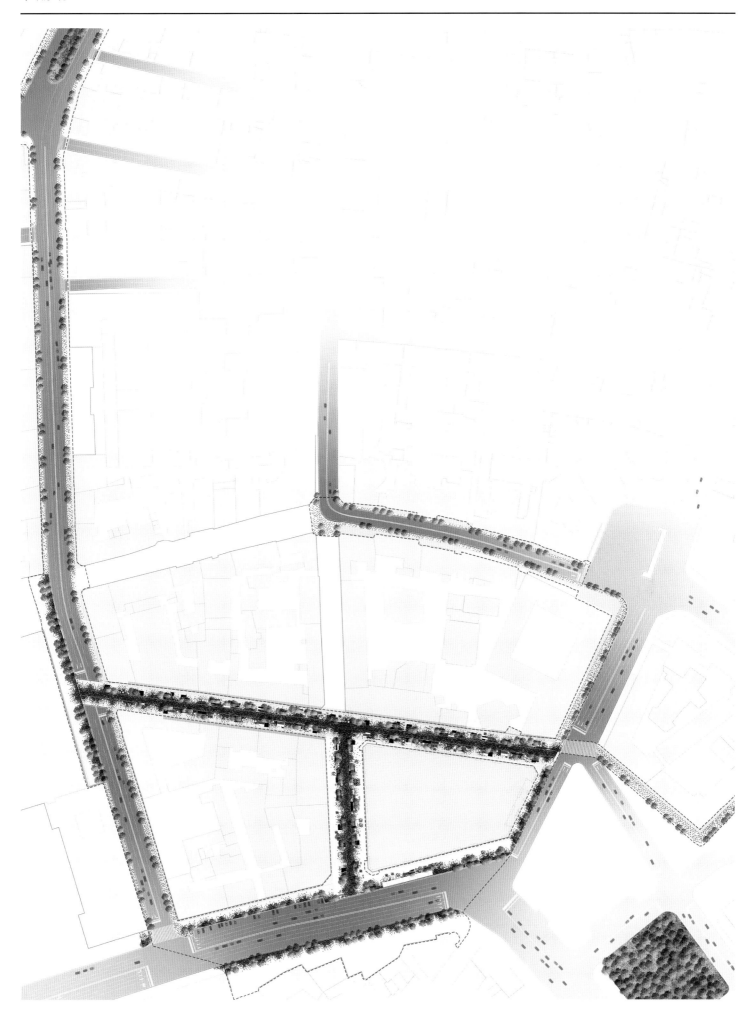

公司办公景观 | **Corporate**

| 公司办公景观 | Corporate |

技术创新中心

美国弗吉尼亚州赫恩登，1988

玛莎·施瓦茨及合伙人设计事务所与 ARQ 合作，对这个技术创新中心进行景观概念化设计。技术创新中心是一个由政府支持的科学研究市场化机构。这块场地面积 0.4 公顷，包括一座上端为梯形的四层车库，车库两侧是一座形状不规则的办公大楼。设计内容主要包括机动车和行人入口、一片小树林以及建筑平台。设计理念是把原本不规则的东西改造成兼具理性和秩序的有机体，并营造"规则式"森林。

出入口处有一个中心花坛，里面种植黄枝山茱萸，交通动线更加顺畅。两块石板作为"欢迎地毯"，引导人们进入办公大楼。两座较小的办公楼上分别有一个平台，上面种植小叶椴树。小椴树林一侧稍微弯曲，下面铺填紫红色的砾石和松散的石块，就像一条花园小路。在小树林内，一排排蓝色球形镜面，看起来就像建筑外立面的镜像，让人联想起树下盛开的鲜花。随意放置的金色和灰色混凝土块构成一个平行四边形广场，与场地内的塔形建筑相呼应。靠近咖啡馆处，由明暗材料铺设的混凝土棋盘可以作为就餐平台。

Center for Innovative Technology

Herndon, VA, USA (1988)

In collaboration with Arquitectonica International, MSP conceptualized and designed the landscape for the Center for Innovative Technology, a government-backed entity that markets scientific research. The 43,000-square-foot site includes the top of a trapezoidal, four-story parking garage flanked by a cluster of irregularly shaped office buildings. Design for the site includes the vehicular and pedestrian entrances, as well as a setting of bosques and terraces for the building. The landscape concept is shaped by the idea of bringing rational order to things that are inherently irrational, making shapes out of living things and creating "formal" forests.

The circularity of the entrance turnaround is reinforced with a central mound planted in yellow-twig dogwood. Two flagstone "welcome mats" lead to the office buildings. A terrace for the smaller of the two office buildings accommodates a bosque of little-leaf linden trees. Gently curved on one side, the bosque is planted in plum-colored crushed gravel with stripes of loose-set stones which create the impression of garden paths. Within the bosque, rows of blue mirror globes appear as pieces of the building's mirrored facade and recall flowers in bloom beneath the trees. Randomly placed gold and grey concrete blocks form a parallelogram-shaped plaza that echoes the shape of the site's tower building. Near the cafeteria, a concrete checkerboard of light and dark pavers forms a dining patio.

公司办公景观 Corporate 223

公司办公景观 Corporate

公司办公景观 Corporate

| 公司办公景观 | Corporate |

贝顿迪肯森公司总部

美国加利福尼亚州圣何塞，1990

项目位于加利福尼亚州圣何塞，主要设计内容包括：把由12个花园组成的空间序列改造成一个两层半高的中庭；把0.8公顷的汽车庭院改造成医学研究户外景观。中庭外围是由木支架组成的绿篱，上搭无花果，里面是倒影池。靠近休息大厅的倒影池，由2.2米×4.9米的绿篱围绕。中庭的另一端同样也有一个倒影池，由0.4米×1.8米的绿篱环绕。绿篱较高且私密性较强的空间作为私人会议场所。墙体较矮的空间，午饭时间可以摆放若干个咖啡座椅。彩色混凝土倒影池旁设置贴有瓷砖的花坛，里面种植虎尾兰。水通过池塘边缘的球形喷头注入池中。

由鱼尾葵组成的柱廊序列对中庭的中轴线进一步起到强化作用。中庭水泥地面为现浇混凝土，搭配绿色和黑色条带。这种格局一直延伸至建筑外墙，对汽车庭院进行有序组织。螺旋状的岩石和造型类似于无花果的穹顶，成为空间的视觉焦点。

Becton Dickenson Headquarters

San Jose, CA, USA (1990)

A graduated series of 12 garden rooms transforms a two-and-a-half-story central atrium and a two-acre motor court for a medical research complex in San Jose, California. Defined by hedges of wooden armatures planted with ficus vines, the rooms enclose reflecting pools and range in size from 24-foot-square spaces bordered by 16-foot-high hedges near the lobby to 4-foot-square by 6-inch-high enclosures at the opposite end of the atrium. The taller, more secluded rooms serve as private conference spaces. The lower-walled spaces offer cafeteria seating at lunch time. Painted concrete reflecting pools surround planters clad with ceramic tile and filled with Sanseveria. Water pours into the pools from globes along the tiled edges of the pools which also provide seating.

A fishtail palm colonnade reinforces the atrium's central axis. The atrium's concrete floor is composed of poured concrete painted in green and black stripes. This pattern extends beyond the building to organize the motor court where a spiraled rock and ficus dome serve as a focal point.

公司办公景观　　　　　　　　　　Corporate

公司办公景观 Corporate 229

公司办公景观 | Corporate | 230

瑞士再保险总部大楼

德国慕尼黑，2002

新建瑞士再保险总部大楼位于慕尼黑附近。这座总部大楼由两座二层高的环形建筑组成，上方是16个小型鞋盒状的办公空间，呈螺旋状向上攀升；下方是一个地下停车场。三层高的走廊包裹着整座大楼，由攀缘植物覆盖。走廊看起来就像一道绿篱，围绕着大楼，悬浮在空中。

项目中，有两块重点景观区域。第一块是围绕环形建筑的绿篱内部，第二块是环形建筑内部的倒影池。地面景观使一系列带条纹的扇形沿着环形的大楼螺旋上升。扇形区域中的条纹与周边农业景观相关联。每个扇形地段采用不同的色彩，有红色、蓝色、黄色和绿色等。每个条带仅由一种植物构成，或者采用某种特殊材料，例如，彩色碎玻璃和彩色砾石。这些条带被步行道所分割。条带中断开的部分方便行人和机动车通行。

其他外部空间可以用来就餐、聚会或者娱乐休闲。扇形区域中设有灯箱，彩色光带从下方车库中放射出来。地面上大楼背后黑暗的阴影，借助于一系列反射材料，例如，镜面和凝视球体，形成一种"光体"。扇形条带在一年四季不同的时间段呈现出不同的色彩。例如，秋季卫矛之"火"熊熊燃烧；春季球茎类植物尽显耀眼的红色；冬季，灌木类植物结满红色的果实。

环形建筑内部的倒影池围合出一个特别适合沉思冥想的小空间，划分为四个部分，并被赋予相应的色彩。关键区域中种植着水生植物，其中有一大片睡莲。倒影池的表面是平静的，但平静的表面下却充满动感与震撼：彩色凝视球、大理石、碎玻璃、砾石以及陶瓷盆等分布于水下，为这个空间注入些许生机和活力。

Swiss Re Headquarters

Munich, Germany (2002)

A new headquarters for the Swiss Re, a German insurance company, is being constructed near Munich, Germany. The headquarters sits atop an underground parking garage and consists of a two-story square doughnut-shaped building with 16 smaller shoe box-shaped office spaces raised above the doughnut and spiraling off of it. Wrapping the entire complex of buildings is an elevated three-story walkway covered in vines. The walkway appears like a hedge floating in space around the building.

The landscape design addresses two primary areas, the zone that lies within the hedge surrounding the doughnut and the reflecting pool in the interior of the doughnut. The design of the ground level is a series of striped quadrants spiraling out from the doughnut. The striping of the quadrants refers to the surrounding agricultural landscape. Each quadrant is a different color: red, blue, yellow, or green. Each stripe is made up of a single plant species or a technical material such as crushed colored glass or colored gravel. Stripes are separated by walkways. Breaks in the stripes allow for a comfortable flow of circulation within and among quadrants.

Other outdoor spaces accommodate dining, meeting, and recreation. Light boxes in each quadrant provide a stripe of colored light glowing from the parking garage below. To deal with the darker shadowy areas under the elevated buildings, the ground is treated as a "volume of light" by using reflective materials such as mirrors and gazing globe. The quadrants are ordered so that the different stripes show their colors at different times during the year. For example, in the red quadrant the Euonymus alatus is a blaze of red in autumn, a field of bulbs blooms with a dazzling red in the spring, and shrubs provide red berries in winter.

The design for the reflecting pool in the interior of the doughnut is conceived of as an elegant, meditative space to catch the light. Like the surrounding landscape, it too is divided into four quadrants with corresponding colors. Water plants are used at crucial areas, one quadrant is filled with water lilies. The pool has a still surface, but just beneath the surface are vibrantly colored gazing globes, marbles, crushed glass, gravel and terra cotta pots that energize what could be an otherwise gloomy, shaded space.

公司办公景观 Corporate 231

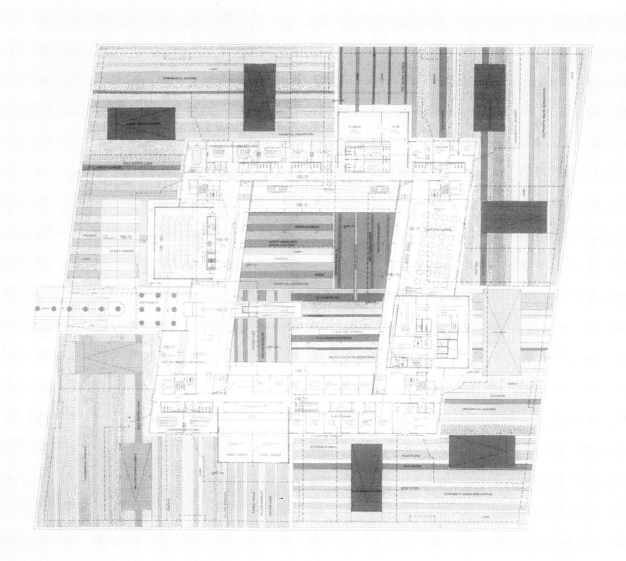

公司办公景观 Corporate

公司办公景观　　　　　　　　　　Corporate　　　　　　　　　　233

公司办公景观　　　　　　　　　　　　　　Corporate

巴克莱银行总部大楼

英国伦敦，2004

巴克莱银行总部大楼位于伦敦金丝雀码头。大楼由HOK建筑设计事务所设计，内部装修由普林高建筑设计事务所设计。这座新建大楼内部五个中庭的绿化和装修由玛莎·施瓦茨及合伙人设计事务所负责。每个中庭六层高，三面由办公室环绕，俯瞰着这块开放空间。

中庭设计的目的是营造氛围独特的办公环境并完善空间的使用功能。每个中庭都是三维立体化的，墙面上悬挂人工材料，地面上种植有机植物。在这六层高的环境当中，每一个空间都是一个独立的办公场所。大楼中的工作人员可以自由选择办公地点及空间的装饰材料。

大楼的第六层，人造竹竿从天花板上以不同的角度悬垂而下，在下面的地板上，用竹子围合成一个圆形座椅休息区。这种设计安排，从空间上方俯瞰，令人感到震撼；从地板上看，既有私密空间，又有户外空间。

在第12层地板上，叶片超乎寻常的喜林芋和龟背竹从天花板上悬垂而下，就像热带丛林。叶片下方种植容器中的龟背竹沿着座椅有序排列。

悬挂在第24层地板上的透明落叶树，植株高大，色彩鲜明，与丛林形成鲜明的对比。这片意象抽象的"落叶森林"与下方地板上的盆栽落叶树构成一种平衡关系。这座大楼的中庭顶部是几何形状比较规则的室内绿篱花园，蓝色和绿色玻璃盒子从天花板上悬垂而下。

每个中庭的设计各不相同。它们组合在一起，构成非正式的休息空间，大楼中的工作人员可以在这里休息和停留。每到夜晚，这些空间在灯光的照耀下呈现出各种不同的形状和图案，映照于大楼内外。

Barclays Bank Headquarters

London, UK (2004)

The new Headquarters for Barclays Bank, designed by HOK Architects with interior outfitting by Pringle Brandon Architects, is located in Canary Wharf, London. Martha Schwartz Partners designed the five interior atria within this new building. Each atrium is six stories high and is surrounded on three sides by offices that look down into the open space.

The design intention for the atria is to create a unique environment within each while allowing for various functions to occur. Each atrium is designed with three-dimensional installations including hanging artificial materials and floor components of organic plant materials. Each of these five spaces creates a unique address for its six-story surroundings. They also provide choices for how and where the building inhabitants want to work within the building.

For the sixth floor installation, artificial bamboo rods hang from the ceiling at varying angles, while below, a circular seating area is delineated on the floor by bamboo plants. This arrangement allows for vibrant views from above and both privacy and an outdoor view at the floor level.

On the 12th floor, oversized artificial philodendron and Monstera deliciosa leaves, hang from the ceiling, creating a tropical, jungle-like canopy. Beneath these leaves, actual Monstera plants sit in planters alongside various seating options.

The 24th floor contrasts the jungle with large, colorful, hanging transparencies of deciduous trees. This abstract deciduous forest is balanced by potted deciduous trees on the floor below. The top atrium of the building takes on the form of a geometric indoor hedge garden with blue and green glass boxes suspended from the ceiling above.

Each atrium is different from the next, yet they all exist as informal breakout spaces, lounges, and gathering points. At night, the spaces are lit to allow the collection of shapes and images to glow on the interior and exterior of the building.

公司办公景观 Corporate

公司办公景观　　　　　　　　　　　　　　　　　Corporate　　　　　　　　　　　　　　　　　　242

万科中心

中国深圳，2013

万科中心是由中国最大的房地产开发公司在深圳建造的一座综合大楼，其长度大约相当于美国帝国大厦的高度。这座大楼内设公寓、办公室、酒店、会议中心、温泉和地下停车场。

玛莎·施瓦茨及合伙人设计事务所负责万科中心的景观设计。景观设计力求对原有景观进行一系列改造与重建，将这里打造成高品质的公共空间和私人领地，并为周边社区居民提供周到的服务。

万科中心的景观设计建立在"群岛"理念的基础上，目的是通过对原有景观元素的精心布置和多种植物的巧妙利用，在以多座小土山为基础的区域环境中营造丰富多元的特色景观。

办公区中的当地草本植物和整齐划一的常绿灌木有助于净化空气。酒店区属于高端景观区，各种装饰植物遍布其中。

另外，万科中心的景观设计还打造了一系列临时性景观及相应的配套设施，例如，季节性花园、儿童室外戏水乐园、游泳池和温泉。

近年来，LEED 金奖认证体系特别关注景观设计的可持续性。万科中心的景观设计巧妙地运用许多可持续性元素和设计方法，包括降水管理和储存、浮岛水分清洁、生境创造以及本土植物和可循环使用的装饰材料等。万科中心就像一个艺术化的城市农场，为区域中的餐馆提供绿色生态的农产品，同时赋予周边社区居民强烈的区域归属感，并让公众了解城市生态知识和食品供应体系。

Vanke Center

Shenzhen, China (2013)

Vanke Center, a mixed use building in Shenzhen by the largest real estate developer in China, is as long as the Empire State Building is tall. It includes apartments, offices and a hotel, with a conference centre, spa and car park below ground level.

Martha Schwartz Partners was appointed as landscape architect to re-envision and improve the existing landscape, and to transform it into high quality public and private spaces, both for the neighborhood surrounding the development and private clientele.

MSP came up with a concept termed "archipelago" that aimed at cleverly maintaining existing structural elements underneath the series of mounds, and employing a variety of planting strategies to diversify the experience of the landscape throughout the development.

The office zone is planted with native grasses and unified evergreen shrubs to improve the sculptural quality of the space. The hotel area was developed as a high-end landscape, with ornamental planting.

The landscape design also introduced seasonal gardens exploring temporality of landscape throughout the year, children's outdoor water play elements and a swimming pool and Spa.

Sustainability was a key driver, particularly because of the development's LEED Platinum rating. The landscape design was developed by a series of sustainable approaches that incorporated storm water management and storage, water cleansing floating islands, native planting, habitat creation and locally manufactured and recycled materials. The development also features a state of the art urban farm for the neighborhood. It will act as a great tool to enhance the sense of community, help educate the public about urban ecology and food systems and supply the grown produce to the on-site restaurant.

公司办公景观 — Corporate

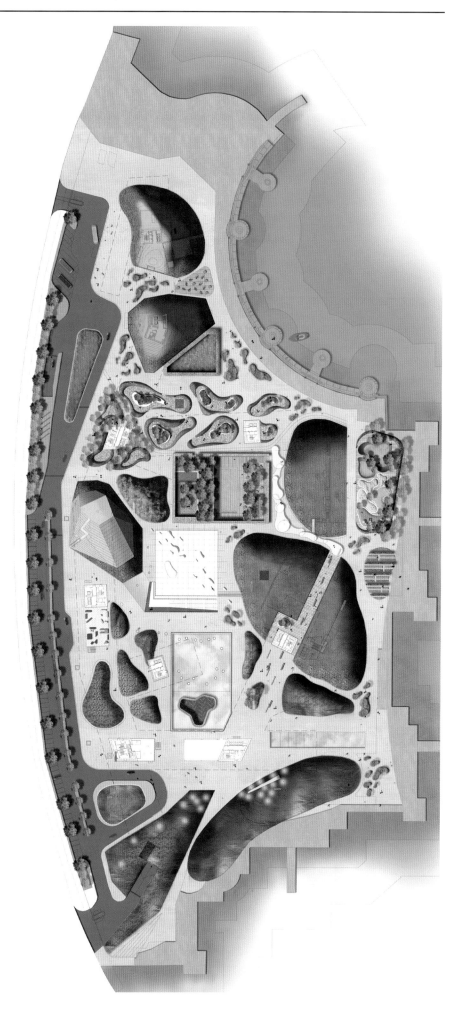

公司办公景观   **Corporate**   244

公司办公景观 | Corporate

Corporate 247

公司办公景观   Corporate

北七家科技商务区

中国北京，2016

项目位于北京市昌平区，北京科技商务区内，是总体规划的第一阶段。面积约为 60 000 平方米。空间类型包括住宅、办公大楼和零售商业用房。

场地的总体设计充分考虑 LEED 金奖认证体系的要求，通过对水的高效利用，降低铺装表面面积，增加绿色覆盖比率，缓解城市热岛效应；充分考虑每个社区的微气候，对冬季西北风设置屏障，促进夏季东南风的流通；在区域南侧设置大型水景，起到降温效果。

景观区域由三个不同的分区或者特征区组成，分别对应的三种不同的用途：商业／零售、中央公园和居住。商业／零售区包括总部办公大楼周边地带、办公区庭院花园、渠北路步行道和生态区。生态区位于场地最北边，是一条具有生态功能的线性景观，场地中的所有降水都在这里汇集和吸收。这块湿地中有休息和散步空间，一个艺术化的入口吸引人们进入绿色心脏地带：中央公园。中央公园是一块开放空间，有"公共绿地"和"下沉花园"。"阳光之角"花园构成下沉草坪区的边界，种植池被抬高，里面种植低矮的绿篱植物、装饰性草坪和多年生草本植物。边缘可供人们坐卧休息或者沐浴阳光、欣赏美景。休闲椅的位置经过精心的设计，并充分利用花园中的阳光直射点。凉风从中央水景吹过，这块海滩洋溢着浓郁的城市气息。

中央公园的另一个主要特征是水景。美丽的水景可供当地居民娱乐休闲，同时在南侧私人住宅区和北侧公共开放空间之间起到隔离作用。

区域南侧的住宅区由小型花园和高大的绿篱植物半围合而成。还有一些私密的景观空间，可供人们沉思冥想。儿童游乐区有许多独特的游乐设施，适合不同年龄的人群。健身房、水景花园以及各种可以供人们娱乐休闲的设施散布于场地之中，有的受到阳光照射，有的处于阴影之中。每一个设计独特的空间都带给人们享受生活、体验生活的乐趣。场地周边还有一条健身小路，可供人们跑步或者进行其他运动。

Beiqijia Technology Business District

Beijing, China (2016)

The project is located in Changping district, Beijing, and belongs to the Beijing Technology Business District and is the first phase of the overall master plan development. The landscape site area is approximately 60,000 square meters. The site is a mixed use development, including residential, offices and retail.

The overall site is designed to consider the LEED Gold accreditation, through effective and efficient use of water, reduction of urban heat island effect by decreasing the amount of paved surfaces and increasing the green ratio, taking into account the microclimate of each zone, by screening the north-westerly winter winds and welcoming the south-easterly summer winds that are cooled further by passing over a large water feature in the south.

The landscape consists of three different zones or character areas, responding to the requirements of each type of programmatic use: Commercial/Retail, Central Park and Residential. The Commercial/Retail area includes the landscapes around the Headquarters Offices, the office courtyard gardens, the Qubei Road Promenade and the Eco Zone Area, which is located at the very north of the site—a linear landscape with an ecological function—collecting and absorbing all the storm water runoff from the site. This mesic habitat also provides room for seating, strolling and one of the two artistic gateway structures, which draw people into the green heart: the Central Park. The Central Park is an open space with the "public green" and the "sunken gardens". Here, a sunny corner garden frames the sunken lawn area with raised planters which are planted with low hedges, ornamental grasses and perennials. Along its edges people can sit and enjoy the sun or lay down on the lounge chairs carefully positioned in the sunny spots of the gardens. The cool breeze from the central water feature will create this beachlike atmosphere in an urban setting.

Another major component of the Central Park is the central water feature, which utilizes treated rainwater to create play opportunities for the local residents and the public. It takes the form of a large curving arc that functions to separate the private residential area in the south from the public open space to the north.

This southern Residential Zone holds small garden rooms, semi-enclosed by tall hedges or feature walls as intimate landscapes for meditation, play areas for children with unique play elements to cater for all age groups, fitness area, gardens with water features and variety of seating elements positioned in sun and shade. Each uniquely designed room celebrates a moment in life. Surrounding the site, a fitness path is also provided as a sports and recreational trail.

公司办公景观　　　　　　　　　　Corporate　　　　　　　　　　249

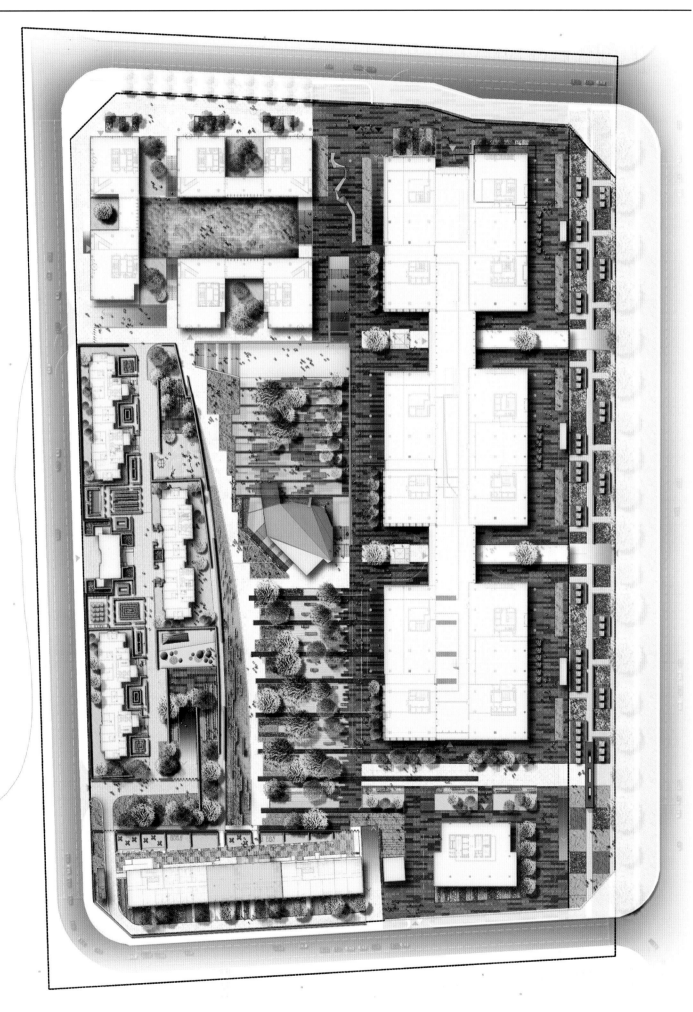

公司办公景观 Corporate 251

公司办公景观 Corporate 252

公司办公景观 Corporate

公园　　　　　　　　　　　　　　　　　　**Parks**

| 公园 | Parks |

滨海线性公园

美国加利福尼亚州圣地亚哥，1988

滨海线性公园对圣地亚哥市的开放区域来说是一个重要的城市贡献。其规模、主题和质量使他成为了一处重要的地标；这个公园就像一件艺术品，在城市与滨水地带之间建立起视觉和文化联系，并使这里交通功能和环境优美地融为一体。

两种格局主导着这块场地：线性铁路工厂和网格状的城市布局。火车、有轨电车及其视觉效果强烈但绝对平坦的轨道是景观设计的重中之重。

滨海线性公园是一个宏伟的公共花园。花园中有各种不同的草地、树木、花坛、苜蓿以及铺满砾石的小路。座椅和绿篱强化了线性铁路院落的视觉效果。轻轨两侧沿线的水面体现了历史上城市与滨水空间关系。在设计上，公园旨在鼓励附近居民随时直接进入公园中慢跑、休闲、晒太阳。一系列步行道网络为居民在此步行创造了有利条件。在公园中，与街道相连的机动车慢行车道整齐划一、连续一致，可以适应未来区域中各种不同类型的开发建设项目，这在圣地亚哥城市景观规划中是独树一帜。

Marina Linear Park

San Diego, CA, USA (1988)

Marina Linear Park was a major civic contribution to San Diego's open space. Its scale, theme, and quality should establish it as an important landmark. The park, as a single work of art, creates a visual and cultural link between the city and the waterfront, integrating the function of transportation into an aesthetic environment.

The two images that dominate the site are the linear dimension of the railroad yards and the grid of the city. The visual strength and absolute flatness of the railroad tracks, along with the historic imagery of train and street cars, are the central themes.

The park is conceived of as a grand public garden. It is composed of rich bands of grass, trees, flower beds, asphalt, and gravel paths. Benches and hedges are dimensioned as abstract reminders of the visual quality of a train yard. Water appears between the LRT lines, referencing the city's historic waterfront relationship. The park is designed to encourage access directly from adjacent housing to accommodate jogging, strolling, and sunning. Pedestrian access is encouraged along a network of pedestrian streets. While development in this area may vary, the streets produce a consistent character of slow moving traffic not found elsewhere in the city.

公园　　Parks　　259

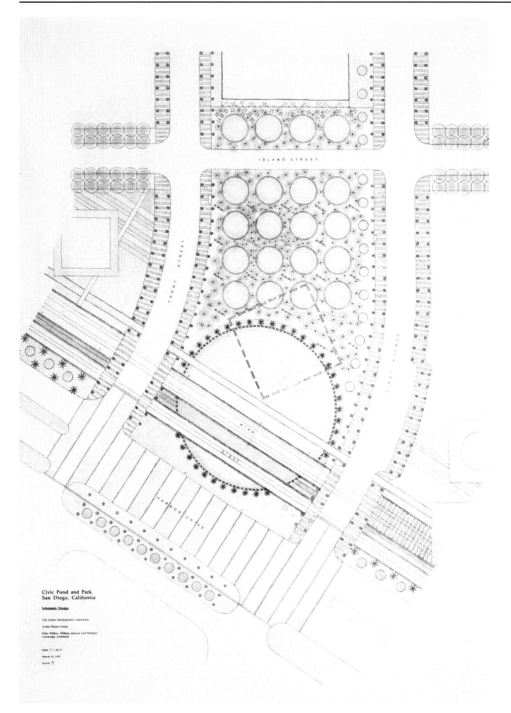

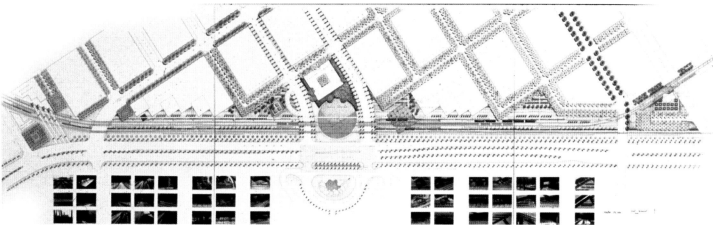

公园 / Parks

约克维尔公园

加拿大多伦多，1995

约克维尔公园的景观设计参照 19 世纪的公园设计实践，把土生土长的景观融入城市之中。这个公园就像一个可以同时存放昆虫、化石和骨骼的传统维多利亚"收藏盒"。运用这个主题，重点关注约克维尔村原始特征和维多利亚风格尺度，将它们最大限度地强化并延伸。

在这个"收藏盒"中，公园被划分成几个宽度各异的区域，植物遍布其中，东段为高地植物群落，中段为低地／湿地植物群落，西段为由耐阴植物构成的花园。主要植物类型和群落包括：松树林、安大略省野生花卉和草坪、杨树林、岩石／草本植物园、海棠树林、湿地、柳絮、边缘草本植物、蕨类植物园。本地植物的巧妙种植有助于延长花园的寿命，随着季节的变换体现生物多样性，并为鸟类和其他动物提供食物和栖息之地。

一大块"空地"是一个步行广场，远离公园的中心，与公园的整体秩序形成鲜明的对比。这块铺装地面可以容纳地铁出入口的大量行人。"空地"的主要特征是，新增加的玻璃包围着地铁出入口，雕塑般的大型岩石让人联想起素有"加拿大保护带"之称的土著基岩。公园中的其他设施包括夏季雨水／冬季冰柱喷泉、便携式咖啡座椅、"遗迹墙"、酸甜乔木等。考虑到夜景效果和安全性，照明设施被重新进行了一番特殊设计。

Village of Yorkville Park

Toronto, Canada (1995)

The landscape design for Yorkville Park is an interpretation of the 19th Century practice of bringing specimens of the native landscape into the city as typified by the tradition of the Victorian "box collections" of insects, fossils, bones, etc. Using this theme, the design reflects, reinforces, and extends the Victorian scale and character of the original Village of Yorkville.

In this "box collection", the park is divided into a series of zones, varying in width and filled with a collection of plant communities ranging from upland communities at the east end, through lowland/wetland communities, to shade gardens at the west end. Major plant types and communities include: pine grove, Ontario wild flowers and grasses, an aspen grove, rock/herb garden, crab apple grove, marsh, Willow "batt", mixed herbaceous border, and fern garden. Native plant material is selected to provide longevity, variety through the changing seasons, and food and habitat for song birds.

A large "clearing", a pedestrian plaza, offset from the center of the park, provides a counterpoint to the general order of the park. This paved area is designed to accommodate the heavy pedestrian traffic in the area of the site's subway entrance. Major features of the "clearing" are a new glass enclosed subway entrance, a large sculptural rock outcropping recalling the native "Canadian Shield" bedrock formation, a summer "rain/winter icicle" fountain, and portable café furniture. Other features of the park include a "relic" wall, a bittersweet arbor, and a variety of seating types. Special lighting is designed for nighttime effect and security.

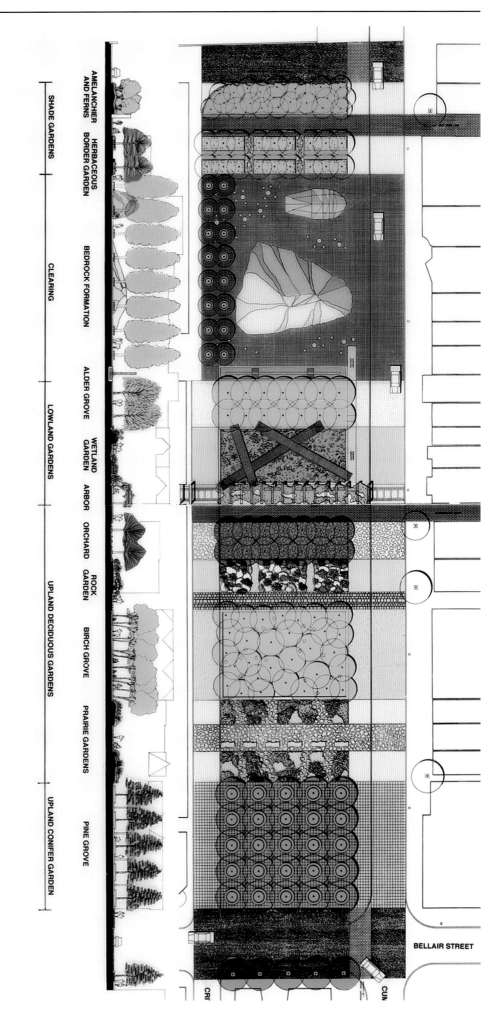

公园 Parks 269

公园   Parks   270

Monte Laa Central Park

Vienna, Austria (2007)

In Monte Laa, a new, very modern part of the city has been created, consisting of offices, apartment blocks and infrastructure. The heart of this district is the central park with its multitude of appearances and functions.

Stretching over the motorway, and located on a former warehouse site, a whole new town is being created, bringing valuable space for living and working. On an area of 90,000 sqm, a total of 200,000 sqm of floor space will be offered for offices and housing, together with a central park covering 12,000 sqm. The basic idea was to connect business premises with residential projects, using the park as a central green between the two. The advantages of the location, together with its ideal connection to the airport and the city, as well as the advantage of the increased residential and recreational value of the site, offer the project increased value. The various thematic elements of the park, grouped around individual platforms and ramps, provide attractive contents for all.

The Central Park is the heart and focal point of the housing project Gartenstadt Laaerberg. The overall premise for the design is to establish the park as a separate strong entity within the project, providing a striking visual identity and an unforgettable experience. Martha Schwartz partners proposed a configuration of elongated sculptural landforms, lines of columnar trees and bands of different materials that work together to emphasize the linear quality of the park. These elements are of the same geometric and formal arrangement as the overall layout of the Gartenstadt.

The most active zone in the park is deliberately located close to the school with an easy connection to the school yard as to invite use by the children. The ramp and ramp surroundings have play facilities for children with the landform paved in asphalt for optimal use by skaters and skate boarders. The end of the landform houses a children's play fountain.

At several locations, Italian Poplar trees are planted along the edge of the park. The design of the housing and office blocks need to accommodate those trees. There is always an obvious connection between the semi-public spaces and the park. The spacing between the trees is wide enough to easily allow access to the park. The buildings at the corners of the park are public space areas that are integrated and form a connection to the semi-public space. The first level of these buildings could accommodate retail shops such as a bank, bakery and pharmacy, or possibly a cafe, to take advantage of the great outdoor space. The entrance road to the office building south of the park is currently under construction and will be roofed.

Parks

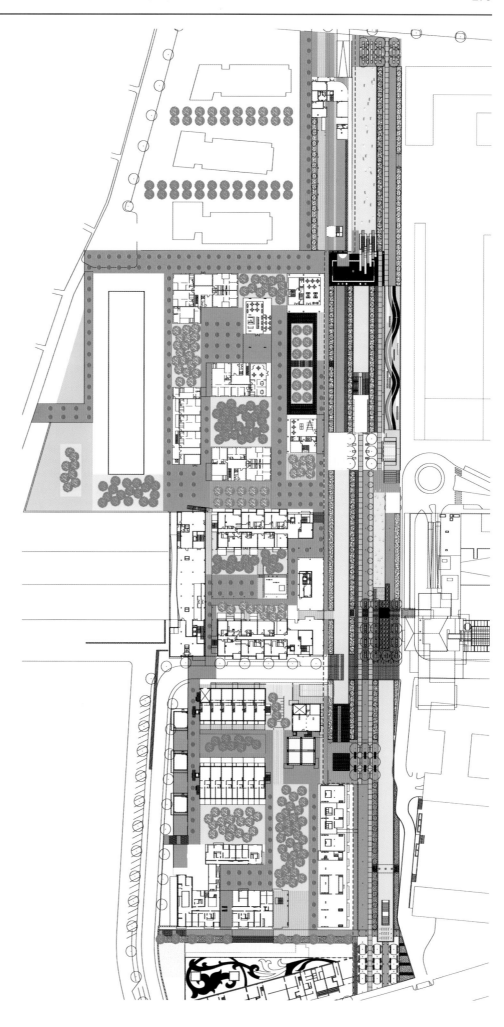

公园　Parks

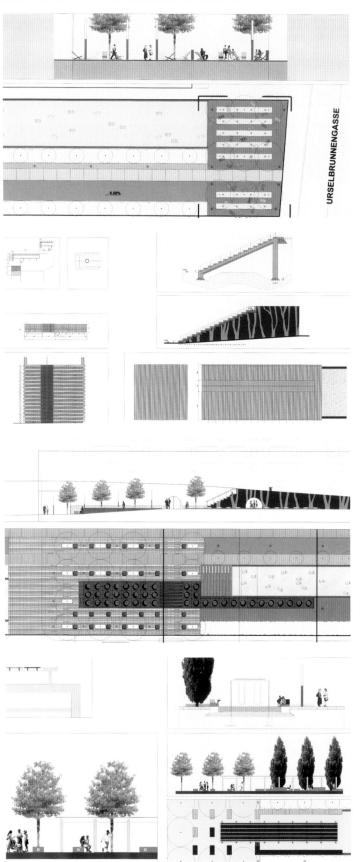
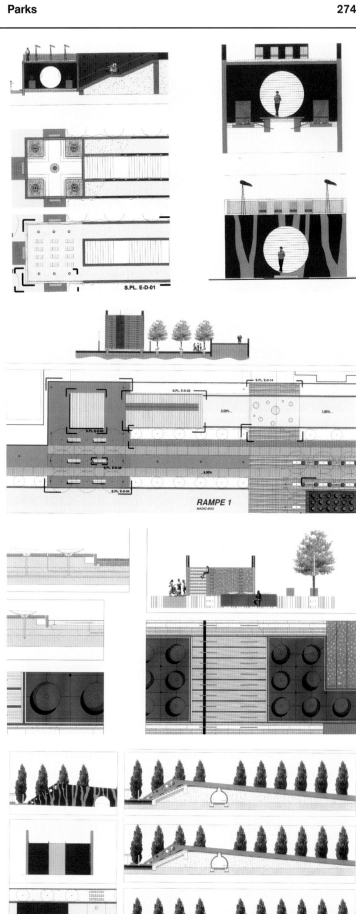

公园　　　　　　　　　　　　　　　　　　　Parks　　　　　　　　　　　　　　　　　　275

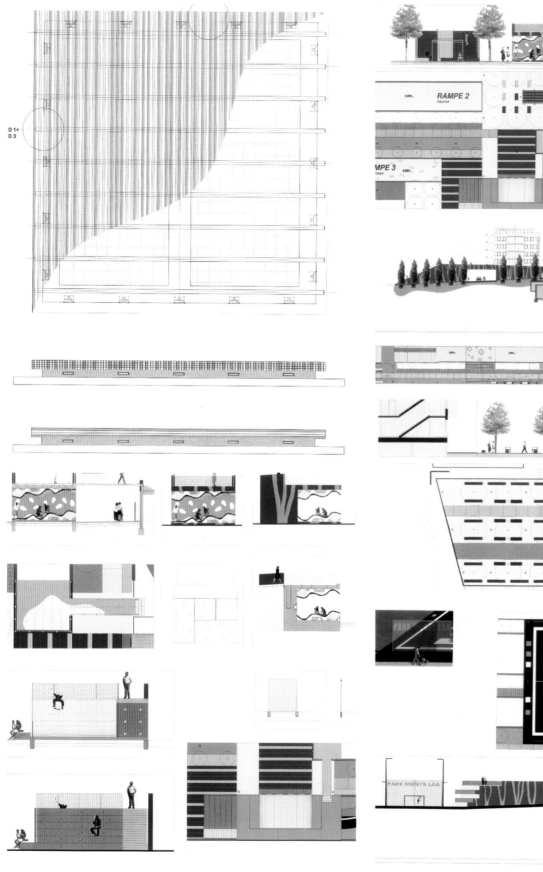
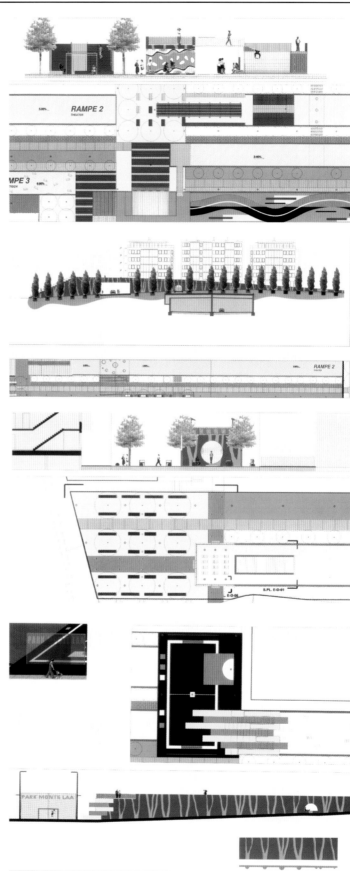

公园 | Parks | 277

公园　　　　　　　　　　　　　　　　　　Parks

公园　　　　　　　　　　　　　　　　Parks　　　　　　　　　　　　　　　　280

公园　　　　　　　　　　　　　　　　　Parks　　　　　　　　　　　　　　　　281

公园 / Parks

圣玛丽教堂公园

英国伦敦，2008

在14世纪，圣玛丽教堂是一个古老的中世纪教区教堂，又名"纽因顿圣玛丽教堂"。1720年，教堂的主要建筑遭到破坏，仅留下历史性地标建筑——钟楼。1721年3月，教堂重建并重新对公众开放。18世纪，教堂被夷为平地，之后被改造成一个公园。

2007年，作为"大象和城堡再生计划"的组成部分，由伦敦开发署牵头提供资金支持，对圣玛丽教堂进行一系列改造。当时，在景观方面，教堂几乎没有留下任何显著的特征。所保留下来的只有教堂边界处的少量墓碑以及经过翻修的栏杆和钟楼上的纪念性石块。

圣玛丽教堂公园的景观设计力求对原有的公园景观进行有序组织，保留具有积极意义的历史性元素，把公园与城市串联在一起，并栽种各种各样的树木、花草，使这个公园成为一个安全、可亲近且令人向往的公共休闲空间。

在现场施工时，工人们发现了一些古老的穹顶和墓穴。经过考古学家的一系列考察和记录，它们留在原地，保持不动。今天，这里仍然是一方神圣的宗教领地，同时也是一片绿色生态之地，供公众娱乐休闲和赏景。

圣玛丽教堂公园的景观设计获得了由英国景观工业协会颁发的"再生景观"优秀奖。

Saint Mary's Churchyard Park

London, UK (2008)

During the fourteenth century, Saint Mary's Churchyard was the site of an old medieval parish church, Saint Mary of Newington. The main structure of the church was demolished in 1720, leaving only the Clock Tower as a historical landmark. The Church was rebuilt and opened in March 1721. During the 18th century, the church was flattened, and the site has been an open park ever since.

In 2007, Southwark Council re-landscaped Saint Mary's Churchyard with the financial assistance of the London Development Agency as part of the Elephant and Castle Regeneration Scheme. At the time of the landscaping works, very little visual evidence remained that the site had been a churchyard. The only remnants were a small number of gravestones placed along the boundary of Churchyard Row, the listed railings which were extensively restored, and the clock tower memorial stone.

The key design goals for Saint Mary's were to reorganize the park while retaining its positive historic qualities, link the park to the city, incorporate healthy existing trees into the design, work with the historic railings, and ultimately to make the park a safe, accessible, and desirable place for the community. These goals are achieved by incorporating park access, safety features, and activity into the design.

While working on the site construction, workmen came across a number of old vaults and burial plots. These were carefully recorded by an archaeologist and left undisturbed. Today the site remains a consecrated ground and continues to be an open space for the use and enjoyment of children and the greater public.

The site has won the British Association of Landscape Industries Award for Regeneration.

Parks

公园 Parks 285

公园 | Parks

凤鸣山公园

中国重庆，2013

凤鸣山公园位于重庆市沙坪坝区，占地 16 000 平方米。公园南侧是老式住宅区，北侧是万科·华誉城，西侧是上桥路，东侧的枫溪路是园区的主入口和最高点。公园于 2013 年春季对外开放。游客穿过公园主路，经过一系列标志性山形雕塑、广场、绿化景观和水景，最后到达万科金城开发销售中心。

该项目力求建造一个示范性公园兼城市公共空间，体现独特性，以促进未来开发项目的销售。公园把游客引向上游主干道上的销售中心和枫溪路主入口，同时与未来开发的邻近地区整合在一起，方便未来的翻修与改造。从停车场到公园最低点的销售中心，这一区域的地形比较极端，行人和机动车的通行面临功能性挑战。然而，另一方面，这里是营造"山形"动态景观的最佳之地。

景观视觉设计力求在设计场地与周围背景之间创造一种强烈的视觉联系。这些周边背景包括：四川盆地中的山峰和谷地、稻田农业景观、长江、弥漫在重庆天空中神秘的大雾。山上亭台、曲折多变的图案、和谐美的地形、各种生动形象的色彩（与蔚蓝色的天空形成鲜明的对比），这些元素都是设计的灵感之源。

枫溪路上的公园入口处，红色和橘黄色的雕塑把人们引向入口广场上的停车场。第一座山亭座落在入口处，作为一系列山峰的起始点，并且沿着斜坡向下延伸。每座山亭的位置经过精心的设计，排列在曲折弯曲的道路两旁，引导行人向山下走去，直到销售中心。这些山亭在白天和夜晚可以提供遮阴，上面的灯光跳跃闪烁，神秘莫测。

蜿蜒曲折的道路使沿路景观不断变化，形成一种地形语言，人们行走其上，就像在一座深山里沿着一条小路向上攀登。每一个拐角处都有一个平台，可供人们休息或赏景。道路两旁的栏杆由大块深黑色混凝土筑成，形态就像一座座岩石，眺望着山坡。墙体之间的谷地缺口形成一道溪流，穿过项目区域。水体是凤鸣山公园的重要组成部分，以"流动"的形式表现出来。从广场到销售中心，有不同的水景，例如，水渠、池塘和喷泉等，它们既可以调节温度，又可以营造迷人的景观氛围。

整个公园构成一个充满欢乐的旅行序列。人们从入口广场的标志性景观，沿着曲折弯曲的小路向下走，穿过蜿蜒曲折的水景，来到广场，进入销售中心。凤鸣山公园已成为重庆市一处重要的城市景观，它深受人们喜爱，令人心驰神往。

Fengming Mountain Park

Chongqing, China (2013)

Fengming Mountain Park is set on a 16,000 square meters site, located in the growing Shapingba District of Chongqing. The site extends south to the old housing quarter, north to the Huayu City Project, west to Shangqiao Road and east to Fengxi Road, which is the main entrance and highest point of the site. Opened in Spring 2013, visitors are taken on a dynamic journey via a series of iconic mountain-shaped follies, plazas, greenery and water features to the proposed Vanke Golden City Development sales centre.

The brief was to design a demonstration park and urban public realm to express a unique identity in order to market the future development. The park is required to draw attention to the development sales centre from the upper main road and entrance from Fengxi Road; and be adaptable for retrofit and integration for the adjacent future development.

Extreme topography creates both a functional challenge to facilitate pedestrian and vehicle movement from the upper carpark to the sales centre at the lowest point; and a unique opportunity to provide a dynamic landscape–the "mountain".

The vision was to create a strong connection between the setting of the site and the surrounding backdrop of the mountainous peaks, valleys of the Sichuan Basin; the agrarian patterning of rice paddy terraces; the Chang Jiang River; the mysterious white/grey misty sky of Chongqing. These elements provide the inspiration for the mountain pavilions, zigzag patterns, orchestrated terrain and the use of vivid colours (to contrast against the sky).

On arrival from Fengxi Road, dancing red and orange sculptures line the entrance to draw people into the arrival plaza carpark. The first of the mountain pavilions stands at the entrance, to mark the start of series of visual mountain peaks, descending down the slope. Each pavilion is strategically positioned along the zigzag path, leading pedestrians down the 'mountain', towards the sales centre area. The pavilions provide shade during the day and at night, are lit to create a spectacular glowing lattern effect.

The zigzag path ensures the extreme level change is accessible for all, the path also becomes a geological pattern language, as if one is a walking on trails winding up a steep mountain. At each turn, a platform provides a place to sit and enjoy the view or take respite from the hill. The zigzag path is lined with balustrade walls constructed from large pieces of deeply textured dark concrete to create a rocky silhouette looking up the hill. The valleys or crevices that are created by the walls become sources for streams which run through the project.

The presence of water is an important part of Fengming Mountain Park and is expressed as a "flow" of water from the arrival plaza to the sales centre, using a variety of different water effects, such as: channels, pools and jets to assist with cooling, provide sounds and atmosphere to what is a captivating landscape. This entire park is a sequence, a triumphant journey, from the patterned markings in the arrival plaza; down the zig zag path; into meandering water features; through the plazas and then on to the final destination at the sales centre. Fengming Mountain has become a vibrant, joyful and well loved part of the Chongqing cityscape.

Parks

公园

287

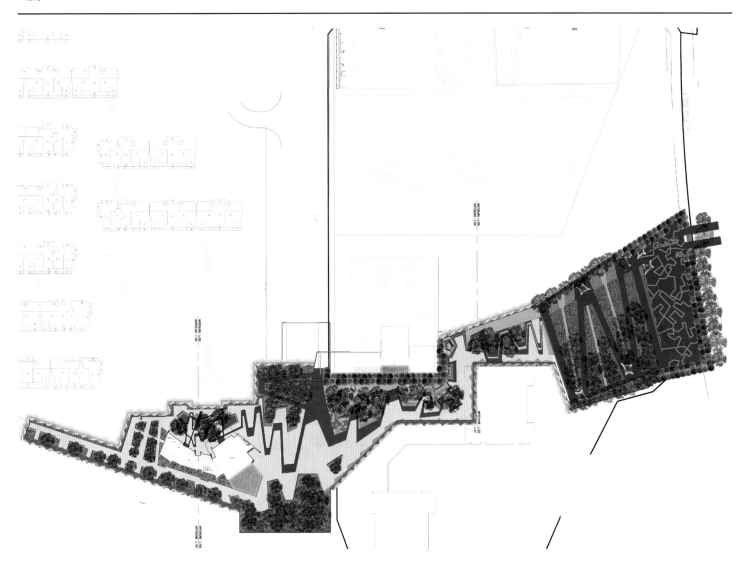

公园　　　　　　　　　　　　　　　　　　　　Parks

公园　　　　　　　　　　　　　　　　　Parks

再生景观 　　　　　　　　　　　　　　　　　　　　Reclamation

温斯洛农场保护景观

美国新泽西州哈蒙顿，1996

这是一个大型农业项目，把艺术、复垦和生态结合在一起。这块麦克尼尔地产位于新泽西州土地贫瘠的松树林之内，面积 243 公顷。这里有丰富多元的自然景观，包括茂密的森林、逐渐起伏的地形、30 公顷废弃的黏土矿、富含矿物质的蓝绿色水域等。同时，这里还可作为社区垃圾场。

项目设计力求对这片污染严重的区域进行一定程度的复垦，使其成为动植物的栖息地，为有机农业开辟一片开阔的田野，为那些对特定地域空间和景观大尺度艺术感兴趣的艺术家提供一方工作场地，并且为麦克尼尔冠军级拉布拉多犬营造一片训练场。

在项目进程中，业主、生态学家、玛莎·施瓦茨及合伙人设计事务所设计团队以及施工团队展开了紧密的合作，深入现场，亲力亲为。项目初期的任务是移除松树，以便开展有机农业。通过精确的测量和有选择地清理废物，场地中多了许多闲置空间。接下来是对场地进行修复，增强地形的起伏感，同时具有雕塑的形式感。再次，对土壤进行改良，将采伐的木材进行修剪后混入土壤中，以增加土壤中的有机质，同时增强无菌黏土的通气性。这种混合土壤有助于促进植物的健康生长。

在美学方面，自然景观、农业景观和文化景观构成一个独特的综合体。自然景观元素与规则化的花园语汇相结合，在未受干扰的景观和更加规则化的花园之间形成关联。公园中的小路和大道经过精心的设计和安排，景色诱人，激起人们的好奇心，渴望进行一番探险。农业大棚和用于贮藏的建筑被改造成画廊和会议空间，并附带花园。

项目设计把那些看起来毫无使用价值的土地和空间相结合：种植有机作物的农田被改造成一个大型花园。经过修剪的整形植物，穿过农田，把农田景观与后巴洛克园整合在一起。复垦后的黏土矿被塑造成超现实主义的形式，野生动物重新回到这片为艺术装置、猎狗训练而营造的景观之中。项目设计为这个衰败凌乱的区域注入了新的生命力；同时，在新泽西州，它在区域文化营造和生态保护方面树立了典范。

Winslow Farm Conservancy

Hammonton, NJ, USA (1996)

This is a large-scale agricultural project that was designed as a marriage between art and the practicalities of reclamation and ecology. The 600 acres McNeil property is an estate located within the New Jersey pine barrens; it contains a diverse range of landscape conditions including dense forests, gradually rolling topography and a 75 acres abandoned clay quarry that holds mineral-rich turquoise water and served as the community dump.

The objective for this project was to reclaim the spoiled and polluted acreage of the clay quarry so that it could once again serve as a habitat for local flora and fauna, to create open fields for organic agriculture, to serve as a retreat for artists who are interested in site-specific, landscape-scaled artworks, and lastly, as a training ground for McNeil's champion Labrador field dogs.

The working process was richly collaborative, on-site and hands-on, working between the client, contractors, ecologists and the MSP design team. Initially, the task was to remove pine trees so to create fields for organic farming. Spaces were carved into the site by calculated, selective clearing. Next, the site was graded to enhance the rolling landscape and create juxtapositions with sculpted forms. Soils were amended by mixing the harvested wood that had been chipped so to incorporate organic matter and to aerate the sterile clay. This mixture would eventually support plant life.

The aesthetics were derived to combine the landscapes of nature, agriculture and culture into a unique mix of these three typologies. Elements of the natural landscape were used in conjunction with formal garden language to create work that posed a dialogue between the undisturbed landscape and more formalized gardens. Paths and roads have been carefully composed and sited so to create vistas and to pique one's curiosity and desire to explore. Agricultural sheds and storage buildings have been transformed into gallery spaces and meeting rooms with attendant gardens.

The composition results in an intriguing combination of unlikely uses and spaces: agricultural fields of organically grown crops are designed as a large-scaled garden. Clipped topiary elements run across these agricultural fields conflating the image of farms with a latter-day baroque garden. The reclaimed quarry is shaped in unlikely surreal forms while wildlife once again inhabits this landscape built for art installations and training hunting dogs. In the end, this site has provided a new life for a once degraded area. Given that there are thousands of such sites in New Jersey, this site has provided a template for cultural and ecological regeneration for others to follow.

再生景观 Reclamation 300

麦克劳德尾矿

加拿大杰拉尔顿，1998

在杰拉尔顿金矿项目中，出于美学和经济原因，需要对尾矿进行重新塑造。安大略省杰拉尔顿金矿位于桑德贝东北大约322千米，是一个已经关闭了的金矿，留下了1400万吨尾矿。这些尾矿遍布于面积69公顷的广大地区，深达8.2米。位于城镇的一个主要路口，这些堆积平坦且面积巨大的尾矿欢迎游客的到来。为了促进经济的再发展，这个城镇决定在这个"尾矿之海"上做点文章，比如，美化尾矿的外观，增强可达性，这将丰富游客的游览体验。

设计方案力求把平坦的尾矿塑造成引人注目的雕塑式地形，使其成为一个具有活力的公路边缘景观和城镇的地标性入口。然而，地形设计并不仅仅限于营造强烈的视觉效果。各种不同的小路可供人们步行、赏鸟、骑山地自行车、滑雪。特别值得一提的是，雪橇滑道的设置促进了这一地区旅游业的发展。

在地形营造方面，技术限制是一个关键要素。型号、体积各异的土方工程设备及其旋转半径为平整的规划设计奠定了基础。这个项目的主要目标是使挖方与填方相互平衡，保证最大土方搬运量不超过15万立方米。挖方深度尽可能最低，因为越往下，砷的含量越高。尾矿底部有盖，上方最大可以增加5米。积水也作为一个设计要素，但是，在重整地形时充分尊重地下水位线。排水系统被保留，拟建的大地艺术景观并不会阻碍视线或影响交通安全。

地面表层增施15～30厘米的泥炭土，以利于植物种植。种植规划主要采用当地乡土草本植物，特别是金色植物。土壤能够支持植物的生长，不需要浇水。对于11号高速公路树木种植规划，总体规划中也做了详细的说明。

麦克劳德尾矿项目彰显了设计的力量，即如何将一块废弃的土地改造成一个美丽与力量并存的大地艺术景观。远胜于一件纯粹的大地艺术品，项目所营造的地形更是一件文化产物，强调了矿业在城镇生活中所扮演的重要角色。

McLeod Tailings

Geraldton, Canada (1998)

In the Geraldton Mine project, mining tailings are being reshaped for both aesthetic and economic reasons. Geraldton, Ontario, located approximately 200 miles northeast of Thunder Bay is the site of a closed gold mine where 14 million tons of tailings from the mines have been left. These tailings cover a 170 acres area of land, 27' deep. This huge flat pile of tailings greets visitors at the main entrance to the town. In order to spur economic redevelopment, the town has made the decision to make something of this sea of tailings. Augmenting their appearance and improving access enhances Geraldton visitors' experience.

Design alternatives sculpt the flat pile into compelling sculptural landforms which become a dynamic roadway edge and a landmark gateway to the town. The landform is designed to be much more than just a powerful visual feature, however. Trails invite one to walk, bird watch, mountain bike, snowboard, or sled. Of special importance to the area's tourism is the inclusion of snowmobile trails.

Technical constraints were key to the final form of the earthwork. The different types and sizes of earth moving equipment and their turning radii provided guidelines for the grading plan. A primary objective in the project was to balance cut and fill, and to maintain a maximum total earth moving of 150,000 cubic meters. Cut is kept to a minimum as arsenic levels are higher toward the bottom of the pile. There is a cap at the bottom of the pile of tailings, and there is a maximum of an additional 5 meters that can occur on top. Standing water was considered as a design element, but the water table has been respected by the re-grading. Storm drainage is maintained and the earthwork will not impede sight lines for traffic safety.

6 to 12 inches of peat topsoil was added to disturbed areas to aid in re-vegetation. A planting plan for the project focuses primarily on native grasses, especially those golden in color. The soil can support plants, although plants will not be watered. The master plan also details tree plantings along Highway 11.

The Geraldton Mine project reveals the power of design to remake a wasteland into a new landscape—a beautiful and powerful earthwork. Even more than an earthwork, this landform is also a cultural artifact, highlighting the location and role of mining in the life of the town.

再生景观　　　　　　　　　　　　　　　　Reclamation　　　　　　　　　　　　　　　　307

总体规划 **Masterplans**

| 总体规划 | Masterplans |

多哈滨海大道设计竞赛

卡塔尔多哈，2003

阿迦汗文化信托公司，代表卡塔尔政府，邀请玛莎·施瓦茨及合伙人设计事务所以及其他七家城市设计、规划、景观设计和建筑方面的国际著名公司参加多哈滨海大道更新改造和总体规划设计竞赛。这条滨海大道，长7.5千米，半月形，八车道。多哈湾的沿岸有良好的管理设施、文化设施、商业设施和公园。竞赛任务书中阐明了这个项目的规划目标："提高海滨大道周边的城市生活质量；通过兼具创新性、文化整合性、环境敏感性的城市规划和景观设计，打造一处国际知名的文化艺术场所，以提升这一地区的国际竞争力。"

玛莎·施瓦茨及合伙人设计事务所将多哈滨海大道的总体规划命名为"白色项链"。在滨水地带，有四个同心半月，即四个"C"。宽阔笔直的滨海大道是行人友好型的；步行道和活动区呈网格状，其间分布着再生湿地和潮间生态区；精心规划的浮动木栈道有助于支持整个海湾的水上出租服务。

四个"C"，把各个颇具竞争力的场地串联起来。位于新建中心公园中央的"十"字轴线、规划中的议会大厦、横贯多哈湾且延伸至大海的新建休闲岛，这些景观元素对四个"C"进一步起到强化作用。

对于其他几个指定的竞赛场地，玛莎·施瓦茨及合伙人设计事务所，也提出了详细的城市设计和景观规划方案，其中包括红树林公园方案和博物馆公园方案。红树林公园位于步行道和木栈道的终点，这里也是设计的一个高潮节点。博物馆公园就像一个多姿多彩的梦幻世界。这里有浮动的花园地毯、木栈道、温室、鸟舍、雕塑花园、蝴蝶房舍、休息室、野餐区、喷泉，以及其他有趣的活动空间和娱乐设施。

宽度、质地各异且变化多样的网格状木栈道将所有景观元素编织在一起。博物馆公园中的红树林花园让人联想到北边木板步行道终点处的红树林公园。然而，在这里，这片红树林并非给人一种主题公园般的游览体验，而是作为景观文化的背景和框景。红树林所营造的空间，具有很强的私密性，亲切宜人，提供遮阴，使棱角分明的网格结构变得柔和。

新建的多哈伊斯兰艺术博物馆坐落在这个公园的边缘，指向花园墙体，形成了木栈道内部"C"字的终点，具有很强的象征意义。这是浮动岛的第一部分，为博物馆公园其他部分的规划设计提供了示范。

Doha Corniche Competition

Doha, Qatar (2003)

The Government of Qatar, advised by the Aga Khan Trust for Culture, invited Martha Schwartz Partners and seven other internationally-known firms of urban designers, planners, landscape architects, and architects to participate in a limited competition for the regeneration and master planning of the Doha Corniche—a 7.5 kilometer (4.7 mls) crescent-shaped, eight-lane highway and belt of prestigious administrative, cultural, and commercial facilities and parks along Doha Bay. The competition's stated intention was to "enhance the quality of urban living along the Corniche and to provide an international cultural and arts identity through innovative, culturally appropriate and environmentally sensitive urban planning and landscaping, highlighted by selected landmark projects of international significance".

The White Necklace, MSP's unified Master Plan for the Corniche, establishes four concentric crescents (the "CS") at the water's edge including: the modulated, pedestrian-friendly Corniche Road; the trellised ("White Necklace") seaside Promenade and activity zone; the regenerated wetland and intertidal Eco-zone; and the sparsely programmed floating Boardwalk, which also supports a water-taxi service throughout the bay.

The "CS" link the eight major urban competition sites and are further defined by the strong, centrally-situated cross axis created by a new Central Park—home to the proposed Parliament building—and the newly-created Pleasure Islands stretching across Doha Bay into open water.

MSP also created detailed urban and landscape designs for a number of designated competition sites including the Mangrove Park and the Museum Park. The Mangrove Park is the culmination and end of both the Promenade and the Boardwalk. The Museum Park is a fantasy world for families and visitors alike comprised of floating carpets of gardens, boardwalks, greenhouses, aviaries, sculpture gardens, butterfly houses, "chill—out" lounges, picnic areas, fountains, and a myriad of other activities.

All these pieces are woven together by a grid of boardwalks varying in width, texture, and design. Within them, the mangrove gardens of the Museum Gardens recall the Mangrove Park that ends the Boardwalk in the north. Here, however, instead of being the subject of the park experience as in the Mangrove Park, the mangroves become the background and frame for a cultural expression of the landscape in which the intimate spaces created by the mangroves provide shade, softness, and relief to the rigor of the grid.

Doha's new Museum of Islamic Arts sits at the edge of this park, poking through the garden wall, creating a significant end-point and destination to the inner "C" of the Boardwalk. It becomes the first of the floating islands and establishes the concept for the rest of the Museum Garden.

总体规划 — Masterplans

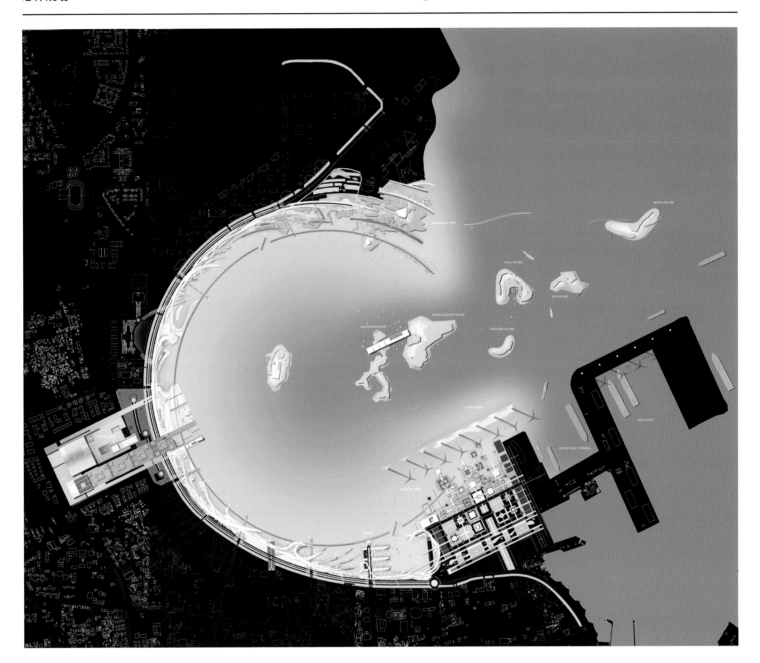

总体规划　　Masterplans　　312

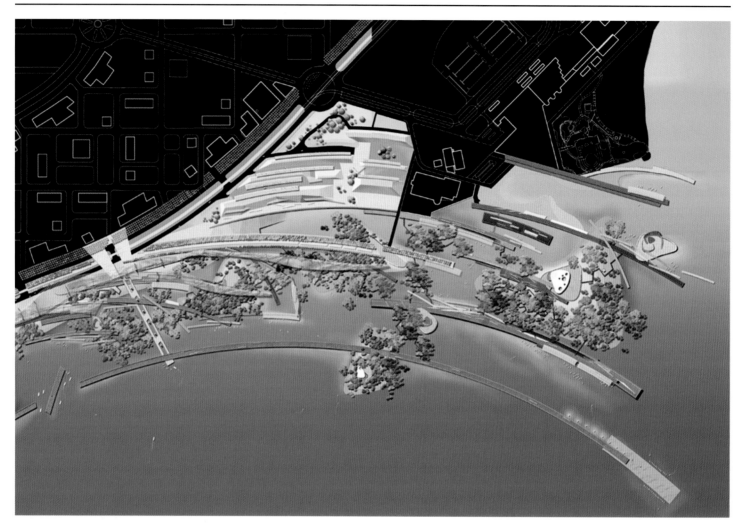

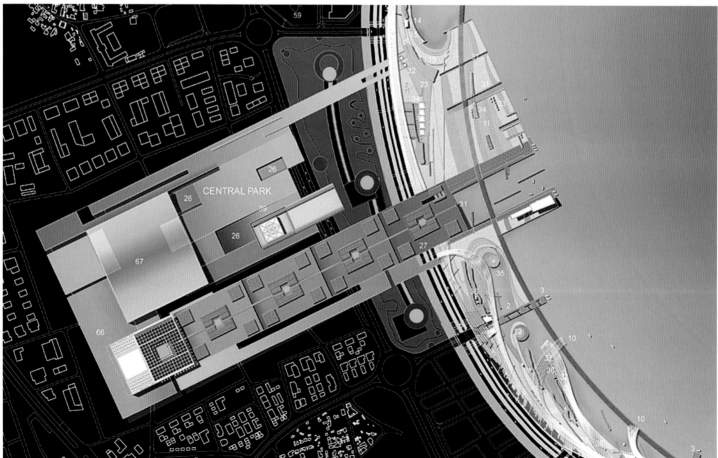

露露岛

阿联酋阿布扎比，2006

玛莎·施瓦茨及合伙人设计事务所与阿魁国际设计公司合作，为阿布扎比露露岛制订了一个总体规划。露露岛景观和开放空间规划力求特色鲜明，并且体现出场地和周边环境特征。景观规划涵盖丰富的景观元素，比如，作为城市动态连接元素的"瓦迪步行道"和大型雕塑般的中央公园将引起全世界对于露露岛的关注，从而丰富岛屿体验，提升自身价值。在"瓦迪步行道"两侧，各种娱乐休闲设施呈组团式布局；各种功能兼容并存，并相互受益。具体的景观和公共空间规划设计力求实现下列目标：

为阿布扎比打造一个新的景点和地标性景观；
打造丰富多彩的空间，吸引人们在这里居住和休闲娱乐；
通过营造宜人的景观环境，提升土地价值，促进旅游业的发展；
尊重自然环境，强调可持续性，把环境和景观摆在重要位置，为中东地区的生活环境树立新的典范；
把高质量的开放空间整合在建筑和基础设施规划之中；
构建更加和谐统一的水陆关系；
建造一个"中央公园"，丰富周边居民的文化生活，为游客提供多元的景观体验；
为阿联酋境内的开发建设项目设定新的标准。

在露露岛上，景观环境及一系列娱乐休闲设施的巧妙营造有助于岛上零售业、宾馆服务业等产业的健康发展，露露岛现已成为当地一个著名的旅游胜地，每年有大批来自世界各地的游客到此旅游、观光、度假。

Lulu Island

Abu Dhabi, UAE (2006)

Martha Schwartz Partners and Arquitectonica International have worked together to produce a master Plan for Lulu Island in Abu Dhabi. The landscape and open space plan for Lulu Island was designed to be distinctive and unique, yet responsive to its site and context. The plan, including elements such as the animated civic connection of the "Wadi Walk" and the large sculptural Central Park bring the attention of the world to Lulu Island and will thus enhance both the experience of the island and its value to the people of Abu Dhabi. The civic, cultural, and recreational programs along the Wadi Walk have been organized into groups or "campuses" of compatible uses so that the diversity of activities can benefit one another. The specific design of the landscape and open space plan is driven by the following objectives:

— create a destination and landmark environment for Abu Dhabi;
— create a variety of enjoyable environments in which to live and recreate;
— create attractive landscape environments that will enhance property value;
— create a new typology for living environments in the Middle East that puts the environment and landscape at the forefront by respecting the natural environment and emphasizing sustainable design;
— integrate quality open space within the architecture and infrastructure programs;
— introduce a more thoroughly integrated land-water relationship;
— provide a "Central Park", a public landscape for all to enjoy that enriches the lives of Lulu Island inhabitants, Abu Dhabi residents, UAE citizens as well as visitors;
— set a new standard for development in the UAE.

The civic, cultural, and recreational components of Lulu Island contribute to the success of the island as a destination, benefiting the island's retail, hotel, and commercial sectors. Branding the island as a civic, cultural, and recreational venue make it a destination for local and regional visitors, as well as for international tourism. Attractive program components also enhance the quality of life on the island, making it a preferred residential location and thereby increasing the value of the island's retail and hotel properties.

总体规划 Masterplans 318

龙山国际商务区

韩国首尔，2012

龙山国际商务区和地标大厦位于韩国首都首尔的心脏地带，具有战略意义和象征意义。东北方是风景如画的南山，西南方是壮观的汉江，东面是宽广的龙山公园。商务区中有一个重要的交通枢纽，串联起区域中的各个组团，共同构成汉江复兴项目。设计方案的重中之重是使商务区与周边环境相呼应，并与周边大片的城市及其腹地有机地串联在一起，使这个项目成为全球经济发展与文化营造的典范。

这个项目与其周边区域及全球环境的联系融入了项目本身，营造了适宜生活、工作和休闲娱乐的环境。各种设施所组成的各个"岛屿"，在保持各自特征的基础上，编织在一起，形成蜿蜒的群岛，作为社区的组成部分，不断地自我强化、自我繁荣。机动车、自行车和人行流线，统一的水系设计，大胆的植物种植策略，把各个分区编织成相互关联的网络空间。

商务区内部包括充满活力的社区和安静的开放空间，汉江沿岸和原铁路线之上的宽敞的公园，极具吸引力，虽然与周边城市区别明显但同时又与之融为一体。效益的最大化可以使项目的发展更持久。每天24小时运行的全球商务总部和文化交通国际枢纽吸引着当地居民和国际游客，带动了地区的复兴和繁荣。宜居性是确保项目在未来实现可持续发展的另一个重要因素。人工湿地、绿色屋顶、太阳能电池板的大量使用以及其他可持续性策略为可持续生活提供支撑，成为景观体验的组成部分，带动了社区的健康发展。然而，这些可持续性策略，只有在积极向上、充满活力的环境中，才能最大限度地发挥自身优势。

Yongsan International Business District

Seoul, South Korea (2012)

The site for the Yongsan International Business District and Landmark Tower is strategically and symbolically located within the heart of Seoul, Korea's historic capital city. Bordered by the picturesque Nam Mountains to the northeast, the majestic Han River to the southwest, and the extensive Yongsan Park to the east, the project also incorporates a major regional transportation hub and represents a central link in the long string of current development plans comprising the Hangang Renaissance Project. Design solutions responding to the inherent cultural significance of surrounding amenities, and to the site's crucial connections to the wider city and region, were essential to the success of this global model for economic and cultural vitality.

The Yongsan project's links to its local, regional and global context filter into the development itself, creating a cohesive environment for life, work, and recreation. Business, Commercial, Residential, Cultural, and Recreational "islands", while maintaining distinct characters, are knitted together into a sinuous archipelago that strengthens and enlivens each of them as part of one vibrant community. Vehicular, bicycle and pedestrian circulation, a unified system of water elements, and a bold planting scheme reinforce the intertwining of the zones themselves into a network of interconnected spaces.

This cohesive community of energetic urban districts and calmer open spaces includes generous parks along the Han River and above the existing railroad, and an alluring formal identity contrasting with the surrounding city yet an integral part of it. The project will achieve much more than a maximization of development profit. Operating 24/7 as a global business headquarters and international hub of culture and communication, attracting residents and visitors worldwide, Yongsan will represent the key component of plans to revitalize the city and region. Equally important, its livability will sustain the development far into the future. Constructed wetlands, green roofs, fields of solar panels, and other strategies for sustainable living will be an integral part of the landscape experience and thus a daily learning experience for the community. However, these sustainability strategies will reach their full potential only in a vigorous, captivating environment.

总体规划 — Masterplans

总体规划 Masterplans

阿布扎比滨海沙滩公园

阿联酋阿布扎比，2012

阿布扎比滨海沙滩公园位于阿布扎比岛西北端。整个项目包括两部分，一部分是原有的沙滩，另一部分是新开垦的沙滩，总长度4千米，由步行道、公园和沙滩组成。在概念设计阶段，沙滩公园与滨海主干道、景观大道、规划中的海洋运输节点以及城市边界之间的界面关系，都是重要的设计元素。

玛莎·施瓦茨及合伙人设计事务所与阿布扎比规划局密切合作，承担了这项设计任务，力求把这块滨海沙滩改造成城市地标景观，作为未来城市可持续发展的典范。设计所面临的关键挑战主要包括：对沙滩和步行道进行合理的组织安排，使其为使用者提供不同的选择，并且满足不同季节和每天不同时间段的使用需求；容纳多种类型的空间（私人与公共），比如，快餐店、餐馆、沙滩俱乐部、自行车道、运动区、儿童游乐区等；强化与城市和大海之间的视觉连接；通过抬高地形，降低繁忙的道路所产生的影响，同时为游客提供良好的观景视野和较为私密的活动空间。

如果游客乘飞机到达阿布扎比，那么这个滨海沙滩是主要的地面标志，构成城市天际线突出的前景。与沙滩公园相连接的景观形态构成独特的城市标识。蜿蜒曲折的步行道一直延伸至沙滩底部，其间穿插着各种各样的小空间。游客可以选择一块空间，进行野餐、日光浴、运动或者阅读。同时，通过有遮阴的景观节点，游客可以一览沙滩广场、运动区和休闲娱乐区的美丽景色，享受一场场奇妙的探险之旅。

Abu Dhabi Corniche Beach

Abu Dhabi, UAE (2012)

The Abu Dhabi Corniche Beach is located on the North West end of Abu Dhabi Island. The project comprises the existing beach and new reclaimed beaches comprising a total of over 4 kilometers length of Promenade, park and sandy beach. The interface with the main Corniche Road, Scenic Boulevard, proposed marine transport nodes, Corniche Park and the edge of the City were important factors to consider during the concept design stage.

Working closely with the Abu Dhabi Urban Planning Council, Martha Schwartz Partners was tasked with restoring the Corniche as the landmark of the city, and as a sustainable model for future generations. Key challenges addressed in the design process included how to organize the beach and promenade so that it provides multiple options for occupation and flexibility for seasonal use and for different times of the day; accommodate the heavily programmed mix including changing rooms, snack bars, restaurants, beach clubs, bike trails, sports areas and children's play areas both for private and public use; maximise the visual connections back into the city and out to the sea, and to mitigate the impact of the busy road using a raised landscape terrain to allow views and provide seclusion for beach users.

The Corniche Beach is the major land mark as you arrive by plane and the prominent foreground to the City skyline. The landscape gesture is important to provide a connection back to the Corniche Park and to create a unique identity for the City. As the total length of beach is over 4 kilometers in length, the design uses a meandering pattern language weaving its way down the stretch of the beach. The "Meander" promenade orchestrates a multiplicity of spaces, rooms and experiences, so that one can claim a space to picnic, sunbath, play and read and still be inside the city and on the beach and takes users on a journey of exploration through a shaded landscape of vantage points to sea and city, leading down to beach-side plazas, sports areas and beaches.

总体规划 Masterplans 325

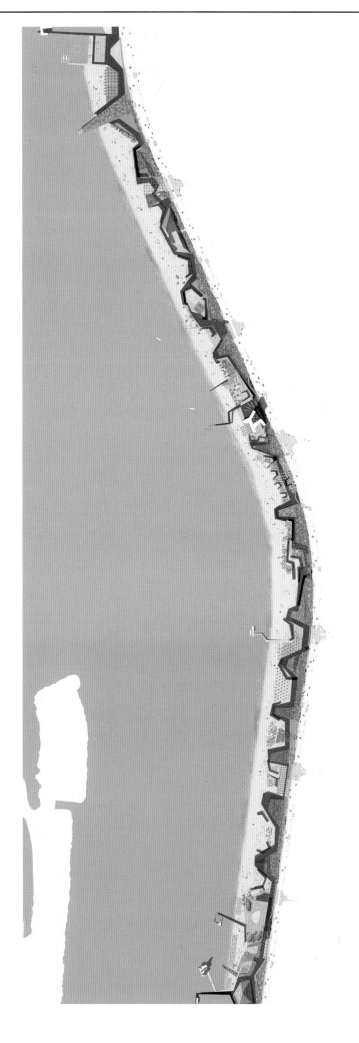

总体规划 / Masterplans

普鲁伊特市规划

印度尼西亚雅加达（规划中）

普鲁伊特市 1 号岛位于历史性城镇中心雅加达北部湾，是一个人工开垦的岛屿，面积 160 公顷，预计容纳 7 万人。项目的总体规划由 SOM 公司制订，其中包括商店、购物广场、办公大楼、公寓大楼和沙滩住宅。在总体规划中，开放空间是一个重要因素，在城市健康和绿色生活方面扮演这关键角色。

一个占地 9 万平方米的大型中央公园把绿色柔和的岛屿西岸与线条硬朗的都市化东岸连接在一起，使这个现代化的中央公园有了服务于整个岛屿，甚至雅加达地区的可能。此外，岛屿上还有一系列相互串联的社区公园，并配有游乐设施、俱乐部、游泳池和运动场地。广场上配有大片的动态的遮阴结构，为便民市场、食品摊位营造舒适宜人的微气候。

长达 6000 米海岸线的滨水区，使得岸边设计成为了该岛屿景观设计的重中之重。设计团队充分利用陆地与海水之间的过渡地带，打造一系列线性的娱乐休闲公园，并配有自行车专用线路和广场。这些景观元素都提供了多样的方式，鼓励和吸引游客去往水边地带。特别值得一提的是，设计团队对现有的大片红树林采取了极富创造性的保护措施，为野生动植物营造了良好的栖息地。

在设计水环境管理体系的过程中，玛莎·施瓦茨及合伙人设计事务所与 SOM 公司、客户密切合作，并充分听取客户的意见。每年 11 月至次年 3 月雨季期间，这里都有大量的降水。为了促进对降水的管理，岛上 5% 的面积配有不同的蓄水区，对降水和灰水进行储存和净化，之后被社区循环利用。这一系列因地制宜的水利基础设施为中央公园 —— 由运河、湿地和滨水游乐区共同构成的综合区域景观的设计提供了独一无二的设计元素和便利设施。

Pluit City

Jakarta, Indonesia (in planning)

Pluit City Island 1 is a proposed 160 hectare reclaimed and constructed island in Jakarta Bay North of the historic town centre anticipated to accommodate a population of 70,000 residents. The project features a masterplan by SOM with Shop Houses (Ruko), shopping plaza, office towers, apartment blocks and beach houses. The open spaces are a prominent factor in the masterplan and play a key role in promoting a healthy urban lifestyle with a green place to live.

A large central park of 90,000 square meters "stitches" together the soft green western edge of the island with the harder and more urban western edge that provides an opportunity for a contemporary park not only for the Island, but for Jakarta as well. In addition, the development provides a number of highly inter-connected community parks, with play areas, club houses, swimming pools and sporting areas. Plazas with extensive and dynamic shade structures offer pleasant micro-climates for markets, food stalls and weekend events.

With an amazing 6,000 meters of waterfront, the treatment of the edge is a major part of the design of the island. MSP's design utilized this transition between land and water to create a series of recreational linear parks, with bicycle circuits, plazas for cafes: all of which employ a range of methods to draw and engage people close to the water's edge. MSP also proposed an innovative approach to extend the protection of existing and creation of new mangrove habitats.

The landscape will have a prominent environmental water management system; each year, during the rainy season between November and March, the volume of water is significant. To help manage this factor, over 5% of the island is occupied by different water bodies to hold and purify storm and grey water that gets recycled for use by the community. This performative water infrastructure becomes a site amenity that introduces a unique element in the design of the Central Park which takes form in a complex landscape program intertwined with canals, wetlands and water play areas.

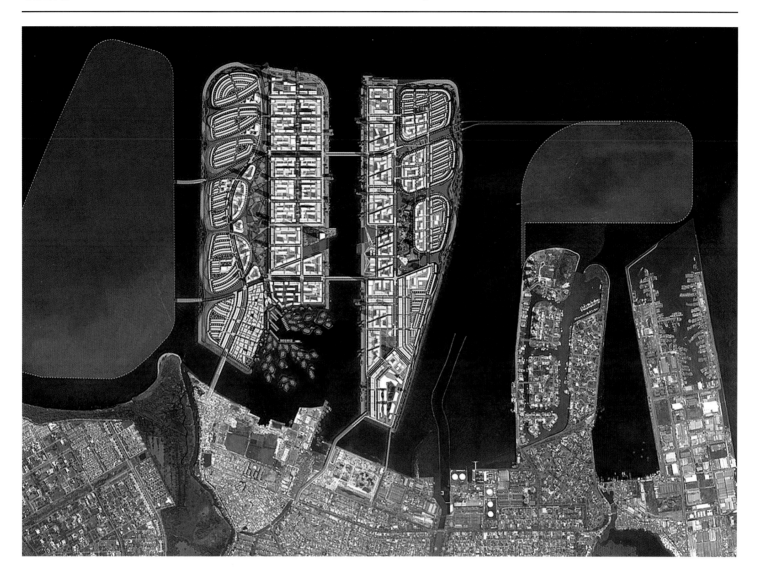

总体规划 | Masterplans | 330

附录 **Appendix**

奖项	Awards

亚太区房地产领袖峰会，最佳未来大项目奖，2014；
美国商业协会最佳剑桥奖，2013；
国际景观设计行业协会设计创新奖，2013；
波士顿风景园林协会优秀设计奖，2012；
美国风景园林学会地标，2012；
美国商业协会最佳剑桥奖，2012；
英国风景园林行业协会更新奖，2008；
芝加哥雅典娜奖最佳环球设计奖，2007；
美国风景园林学会荣誉奖，2007；
库珀-休伊特国家设计奖，2006；
城市土地研究院优秀设计奖，2006；
杰出女设计师奖，美国波士顿建筑师协会，2005；
娱乐休闲奖，2003；
新美国规划协会新墨西哥宪章创新型交通规划奖，2001；
美国风景园林学会优秀设计奖，2000；
美国国家艺术基金会联邦设计成就奖，2000；
美国风景园林学会优秀设计奖，1999；
布劳沃德县公共艺术奖，1998；
美国总务管理局国家设计奖，1998；
纽约公园协会菲利普 N. 温斯洛景观设计奖，1997；
多伦多市多伦多城市设计奖，1997；
美国风景园林学会荣誉奖，1997；
美国总务管理局国家设计奖，1997；
戴德地铁公共艺术设计奖，1996；
美国风景园林学会荣誉奖，1991；
美国风景园林学会优秀设计奖，1991；
美国风景园林学会荣誉奖，1989；
美国风景园林学会优秀设计奖，1989；
美国风景园林学会优秀设计奖，1988；
亚特兰大城市设计委员会奖，1987。

MIPIM Asia Award, Best Future Mega Project, 2014;
U.S. Commerce Association Best of Cambridge Award, 2013;
International Landscape Design Industry Association Design Innovation Award, 2013;
Boston Society of Landscape Architects Merit Award, 2012;
American Society of Landscape Architects Landmark Award, 2012;
U.S. Commerce Association Best of Cambridge Award, 2012;
British Association of Landscape Industries Regeneration Category Award, 2008;
Chicago Athenaeum Award for Best New Global Design, 2007;
American Society of Landscape Architects Honor Award, 2007;
Cooper-Hewitt National Design Award, 2006;
Urban Land Institute Award for Excellence, 2006;
Boston Society of Architects Women in Design Award for Excellence, 2005;
Play and Leisure Award, 2003;
New American Planning Association Award for Innovative Transportation Planning, 2001;
American Society of Landscape Architects Merit Award, 2000;
Federal Design Achievement Award, National Endowment for the Arts, 2000;
American Society of Landscape Architects Merit Award, 1999;
Broward County Public Art Percent for Art Award, 1998;
General Service Administration Design Awards, National Design Citation, 1998;
The Parks Council New York, Phillip N. Winslow Landscape Design Award, 1997;
City of Toronto, Toronto Urban Design Award, 1997;
American Society of Landscape Architects Honor Award, 1997;
General Service Administration Design Awards, National Design Citation, 1997;
Metro-Dade Art in Public Places Award, 1996;
American Society of Landscape Architects Honor Award, 1991;
American Society of Landscape Architects Merit Award, 1991;
American Society of Landscape Architects Honor Award, 1989;
American Society of Landscape Architects Merit Award, 1989;
American Society of Landscape Architects Merit Award, 1988;
Atlanta Urban Design Commission Award, 1987.

设计团队 Team

2016

Hannin Alnimri
Naomi-Sian Bailey
David Baker
Ceylan Belek Ombregt
Alberto Campagnoli
Francisco Coll Carreras
Bo Cui
Sellua Di Cegli
Xinli Du
Richard Hawks
Alicia Hidalgo Lopez
Markus Jatsch
Linli Jiang
Rajiv Kale
Annette Kastka
Edith Katz
Diego Pacheco Gonzalez
Yoon Joo Lee
Eleni Oureilidou
Claudia Pandasi
Jin Koo Park
Kyung Eui Park
Martha Schwartz
Eike Selby
Jingbin Song
Gabriel Tanase
Colin Varley
Ibrahim Diaz Vera
Charlotte Wilberforce
Matthew Williams

设计师小传　　　　　　　　　　　　　　　　　　　　Biographies

玛莎·施瓦茨

理学博士
罗马美国学院院士
美国风景园林师学会会员
英国皇家建筑师学会荣誉院士
英国工业设计师学会荣誉会员

玛莎·施瓦茨，景观设计师、艺术家，主要从事城市景观设计。施瓦茨作为玛莎·施瓦茨及合伙人设计事务所的负责人，在景观设计和艺术设计行业有超过35年的从业经历。其作品涉及的范围非常广泛，遍及全世界，并与许多世界著名的建筑设计师进行了卓有成效的合作。

施瓦茨因其杰出的成就，荣获了众多奖项与称号，其中包括：
皇家工业设计师荣誉奖（由英国皇家艺术、制造和商业促进学会颁发）；
美国库柏-休伊特国家建筑设计奖；
杰出女设计师奖（由美国波士顿建筑师协会颁发）；
北爱尔兰贝尔法斯特阿尔斯特大学荣誉理学博士学位；
美国城市设计研究院研究员；
拉德克利夫学院和罗马美国学院研究员；
英国皇家建筑师学会荣誉院士；
荣誉设计奖（由美国风景园林师学会颁发）。

施瓦茨是哈佛大学研究生院景观设计学终身教授、哈佛大学可持续性城市工作组创始人。关于"可持续性城市和城市景观"这一主题，她在世界上许多国家和地区做过多次演讲。她的作品被广泛出版，或被艺术馆收藏。

Martha Schwartz

DSc, FASLA, Hon FRIBA, Hon RDI, RAAR
Principal

Martha Schwartz is a landscape architect and artist with major interests in cities, communities and the urban landscape. As principal of Martha Schwartz Partners, she has over 35 years of experience as a landscape architect, urbanist and artist on a wide variety of projects located around the world with a variety of world-renowned architects.

She is the recipient of numerous awards and prizes including the Honorary Royal Designer for Industry Award from the Royal Society for the Encouragement of Arts, Manufactures and Commerce for her outstanding contribution to UK design; the Cooper Hewitt National Design Award; the Women in Design Award for Excellence from the Boston Society of Architects; a Doctor of Science from the University of Ulster in Belfast, Ireland; a fellowship from the Urban Design Institute; visiting residencies at Radcliffe College and the American Academy in Rome; an Honorary Fellowship from the Royal Institute of British Architects and most recently a Council of Fellows Award by the American Society of Landscape Architects.

Martha Schwartz is a tenured Professor in Practice of Landscape Architecture at the Harvard University Graduate School of Design and a founding member of the Working Group of Sustainable Cities at Harvard University. She has lectured both nationally and internationally about sustainable cities and the urban landscape, and her work has been featured widely in publications as well as gallery exhibitions.

| 设计师小传 | Biographies |

马库斯·詹斯奇

博士
英国皇家建筑师学会会员
英国皇家艺术协会会员
合作伙伴

马库斯·詹斯奇，建筑设计师和城市规划师，拥有风景园林和艺术学背景，荣获众多奖项；专业设计经历超过 20 年，参与了玛莎·施瓦茨及合伙人设计事务所的所有项目设计，并负责研究工作；与玛莎·施瓦茨在项目设计上的合作超过 15 年；在加入玛莎·施瓦茨及合伙人设计事务所之前，在大卫·奇普菲尔德建筑设计公司工作。

詹斯奇经常在国际性的设计学院从事教学工作，最近主要是在哈佛大学研究生院教授城市设计学，同时，作为维也纳应用艺术大学访问学者；于 2004 年出版《无界空间：空间视觉感的不确定性》一书（出版地：德国斯图加特，2004 年）；并定期就可持续性城市和公共空间设计这一主题发表演讲。他的作品在世界范围内被广泛出版，也常见于博物馆和艺术展览馆。

詹斯奇曾多次荣获大奖和奖学金资助，比如，富布赖特委员会奖、德国学术交流中心研究生奖学金等。同时，他也是伊斯灵顿伦敦自治市设计评估委员会成员以及英国皇家艺术、制造和商业促进学会院士。

Dr Markus Jatsch

RIBA, FRSA
Partner

Markus Jatsch is an award winning architect and urban planner with a background in landscape architecture and fine arts. He has over 20 years of professional experience and is involved with the design of all projects at MSP where he also oversees all research activities. He has collaborated with Martha Schwartz on selected projects since over 15 years and has worked with David Chipperfield Architects before joining MSP.

Markus Jatsch teaches frequently at international design schools, most recently at the Harvard University Graduate School of Design and as visiting professor at the University of Applied Arts in Vienna. He is the author of *Debordered Space: Indeterminacy within the Visual Perception of Space* (Stuttgart 2004) and he lectures regularly on sustainable cities and the public realm. His work has been widely featured internationally in publications as well as in museum and gallery exhibitions.

Markus Jatsch is the recipient of several fellowships and grants such as the Fulbright Commission Award and the German Academic Exchange Service Postgraduate Scholarship. He is a member of the London Borough of Islington Design Review Panel and an elected Fellow of the Royal Society for the Encouragement of Arts, Manufactures and Commerce.

设计师小传 | Biographies

埃克·塞尔比

合作伙伴

埃克·塞尔比，拥有17年的专业设计实践，设计经历丰富，包括从最初的与客户的接触与交流，到概念性设计和总体设计的进一步发展，再到细部设计以及多种项目类型和预算投标文件的制作等，设计对象涵盖高端住宅区、商业和工业开发项目等。

塞尔比在接受了长时间的景观设计专业培训之后，获得了建筑管理学硕士学位，其设计作品主要集中在中东和北非地区。在迪拜，他作为大型项目的项目经理，带领团队，负责棕榈岛和皇家宫殿项目的景观施工，长达11年。他的管理才能得到最大限度的发挥；他工作积极，在时间和预算上对项目及时追踪，使复杂的大型项目顺利落成。此外，他拥有较强的现场施工指挥能力，监督引导项目的现场施工，确保设计方案付诸实施，实现预定的目标。

在玛莎·施瓦茨及合伙人设计事务所，塞尔比主要负责各种项目的日常运行，包括合同管理、方案管理和项目管理。

Eike Selby

Partner

With over 17 years of professional practice Eike Selby has experience in all aspects of the design process from initial client contact through to concept and design development, detail design and tender package documentation on a wide range of project types and budgets including high-end residential, commercial and industrial developments.

Originally he trained as a landscape architect and then received the masters in construction management that led to a career trajectory focused in the Middle East and North Africa region. Working for eleven years on-site dealing with implementation of landscape schemes as project manager of some large-scale projects in Dubai, he led teams for the Palm Jumeirah and palace complexes for royal families. He brings his management skills to making the work flow at MSP proceed grounded, on track for time and budget with international clients, who bring large, intricate projects with ambitious goals. His technical skills, learned from on-site construction help to keep the studio projects realistic and achievable.

Eike oversees the operations of MSP, including contracts and proposals and project management.

夏洛特·威尔伯福斯

合作伙伴

夏洛特·威尔伯福斯拥有企业管理的相关背景，在营销传播行业从业超过17年；除景观设计行业之外，在医疗、电信、媒体及时尚等行业也有所涉猎，设计作品遍布欧洲、亚洲和北美洲。

威尔伯福斯主要从事公共管理工作，并已获得了一些公众人物的支持，包括国际政治家、社会名流和商界领袖。她所出版的营销材料，既有印刷形式的，也有数字形式的；很多歌唱家、谱曲者和生产商为她组织举办的营销活动创作音乐和视频。

威尔伯福斯创办了一家慈善机构，并且运营良好。该机构秉承人道主义原则，斥责国际人口贩卖，引发了环球媒体的广泛关注，包括英国公司（BBC）以及终结剥削和杜绝人口贩卖宣传影片（MTV Exit）。对慈善活动的热衷和对人道主义原则的坚守在一定程度上丰富并推动了威尔伯福斯的景观设计创作，使其作品饱含浓郁深刻的人文关怀。

在玛莎·施瓦茨及合伙人设计事务所，威尔伯福斯负责企业运营和财务方面的日常事务，包括商业发展与对外交流。

Charlotte Wilberforce

Partner

Charlotte Wilberforce has a background in business administration and marketing communications with over 17 years of experience. In addition to landscape architecture, she has worked in medical, telecommunications, media and fashion, spanning Europe, Asia and North America.

Charlotte has directly engaged with and gained the support of public figures including politicians, celebrities and business leaders internationally. She has published marketing materials both in print and digital forms and has inspired singers, songwriters and producers to create music and videos for her marketing campaigns.

Having set up and successfully run her own charity; campaigning to raise awareness of human trafficking, she has gained the attention of the global media including the BBC and MTV Exit. Her philanthropic and humanitarian activities are balanced by a love of art and design and their commercial applications.

Charlotte oversees the business and financial side of MSP, including business development and communications.

| 参与人员名单 | Credits |

阿布扎比滨海沙滩公园
地点：阿联酋阿布扎比
客户：阿布扎比市规划局
年份：2012
状态：停止
设计团队：玛莎·施瓦茨、马修·盖驰、马库斯·詹斯奇、彼得·皮特、德博拉·纳甘、纳吉尔·考赫、杰兰·贝莱克、马蒂亚·加姆巴德拉、克里斯·王、丽贝卡·奥尔、弗劳恩·施瓦茨、里贝克·奥尔、伊丽莎白·莱迪
建筑设计：詹斯奇－劳克斯建筑设计公司

铝酸盐景观
地点：冰岛雷克雅未克
客户：雷克雅未克艺术博物馆
年份：2008
状态：已完成
规模：200平方米
设计团队：玛莎·施瓦茨、艾利森·戴利、雅各布·沃克、丹·盖斯

面包圈花园
地点：美国马萨诸塞州波士顿
客户：玛莎·施瓦茨
年份：1979
状态：已完成
设计团队：玛莎·施瓦茨
摄影：艾伦·沃德

巴克莱银行总部大楼
地点：英国伦敦
客户：巴克莱银行
年份：2004
状态：已完成
设计团队：玛莎·施瓦茨、约翰·佩格
建筑设计：HOK建筑设计事务所、普林高建筑设计事务所
摄影：克里斯·加斯科因、莫利·冯·斯滕伯格

贝顿迪肯森公司总部
地点：美国加利福尼亚州圣何塞
客户：贝顿迪肯森公司免疫细胞计量分部
年份：1990
状态：已完成
景观设计：玛莎·施瓦茨工作室、肯·史密斯工作室、大卫·迈耶工作室
设计团队：玛莎·施瓦茨、肯·史密斯、大卫·迈耶、道格·芬德利、大卫·庄、莎拉·费奇德
建筑设计：金斯勒建筑设计事务所
奖项：美国风景园林学会优秀设计奖，1991

贝鲁特滨水公园设计竞赛
地点：黎巴嫩贝鲁特
客户：索里迪公司
年份：2011
设计团队：玛莎·施瓦茨、马库斯·詹斯奇、Soojung Rhee

Abu Dhabi Corniche Beach
Location: Abu Dhabi, UAE
Client: Abu Dhabi Urban Planning Council
Year: 2012
Status: Project stopped
Design Team: Martha Schwartz, Matthew Getch, Markus Jatsch, Peter Piet, Deborah Nagan, Nigel Koch, Ceylan Belek, Mattia Gambardella, Chris Wong, Rebecca Orr, Faun Schwartz, Elizabeth Leidy
Architects: Jatsch Laux Architects

Aluminati
Location: Reykjavik, Iceland
Client: Reykjavik Art Museum
Year: 2008
Status: Completed
Design Team: Martha Schwartz, Allison Dailey, Jacob Walker, Dan Gass

Bagel Garden
Location: Boston, MA, USA
Client: Martha Schwartz
Year: 1979
Status: Completed
Design Team: Martha Schwartz
Photographer: Alan Ward

Barclays Bank Headquarters
Location: London, UK
Client: Barclays Bank
Year: 2004
Status: Completed
Design Team: Martha Schwartz, John Pegg
Architect: HOK Architects, Pringle Brandon Architects
Photographer: Chris Gascoigne, Morley von Sternberg

Becton Dickenson Headquarters
Location: San Jose, CA, USA
Client: Becton Dickinson Immunocytometry Division
Year: 1990
Status: Completed
Landscape Architect: The Office of Martha Schwartz, Ken Smith, David Meyer
Design Team: Martha Schwartz, Ken Smith, David Meyer, Doug Findlay, David Juang, Sara Fairchild
Architect: Gensler and Associates
Award: ASLA Merit Award 1991

Beirut Waterfront Park Competition
Location: Beirut, Lebanon
Client: Solidere
Year: 2011
Design Team: Martha Schwartz, Markus Jatsch, Soojung Rhee

| 参与人员名单 | Credits | 343 |

北七家科技商务区
地点：中国北京
客户：北京宁科房地产开发公司
年份：2016
状态：已完成
设计团队：玛莎·施瓦茨、马修·盖驰、马库斯·詹斯奇、安妮特·卡特卡、丹·雷亚、伊格纳西奥·洛佩斯-布森、雅斯敏、昂格、艾丽西亚·伊达尔戈、洛佩兹、崔波、迭戈·帕切科、冈萨雷斯、弗朗西斯卡·诺伊斯·弗特拉、吉列斯·得·威弗、胡安·加西亚·格雷罗、纳吉尔·考赫、伊迪丝·卡茨、李奇迪、Kyung Eui Park
建筑设计：RTKL建筑设计事务所
摄影：泰伦斯·张

布劳沃德县市民体育场
地点：美国佛罗里达州福特劳德代尔
客户：阿雷那开发公司
年份：1998
状态：已完成
设计团队：玛莎·施瓦茨、李泰尔·费边、詹姆斯·劳德、特里西娅·贝尔斯
摄影：玛莎·施瓦茨
获奖：布劳沃德县公共艺术奖，1998

技术创新中心
地点：美国弗吉尼亚州赫恩登
客户：技术创新中心
年份：1988
状态：已完成
景观设计：彼得·沃克工作室、玛莎·施瓦茨工作室
设计团队：玛莎·施瓦茨、肯·史密斯、大卫·迈耶、马丁·普瓦里埃

城堡购物中心
地点：美国加利福尼亚州考莫斯
客户：特拉梅尔公司
年份：1991
状态：已完成
景观设计：彼得·沃克工作室、玛莎·施瓦茨工作室
设计团队：玛莎·施瓦茨、肯·史密斯、大卫·迈耶
奖项：美国风景园林师协会荣誉奖，1991

"城市与自然"主题花园
地点：中国西安
客户：西安国际园艺博览会组委会
年份：2011
状态：已完成
设计团队：玛莎·施瓦茨、唐纳德·夏普、艾利森·戴利、克丽丝特·李
摄影：王根

明日之城：Bo01住宅示范区
地点：瑞典马尔默
客户：欧洲住宅博览会
年份：2001
状态：已完成
设计团队：玛莎·施瓦茨、弗朗西·科米尔

Beiqijia Technology Business District
Location: Beijing, China
Client: Beijing Ningke Real Estate
Year: 2016
Status: Completed
Design Team: Martha Schwartz, Matthew Getch, Markus Jatsch, Annette Kastka, Dan Rea, Ignacio Lopez-Buson, Jasmine Ong, Alicia Hidalgo Lopez, Bo Cui, Diego Pacheco Gonzalez, Francisca Neus Frontera, Gilles de Wever, Juan Guerrero Garcia, Nigel Koch, Edith Katz, Qidi Li, Kyung Eui Park
Architect: RTKL
Photographer: Terrence Zhang

Broward County Civic Arena
Location: Fort Lauderdale, FL, USA
Client: Arena Development Company
Year: 1998
Status: Completed
Design Team: Martha Schwartz, Lital Fabian, James Lord, Tricia Bales
Photographer: Martha Schwartz
Award: Broward County Public Art Percent for Art Award 1998

Center for Innovative Technology
Location: Herndon, VA, USA
Client: Center for Innovative Technology
Year: 1988
Status: Completed
Landscape Architect: The Office of Peter Walker and Martha Schwartz
Design Team: Martha Schwartz, Ken Smith, David Meyer, Martin Poirier

Citadel Shopping Center
Location: Commerce, CA, USA
Client: Trammel Crow Company
Year: 1991
Status: Completed
Landscape Architect: The Office of Peter Walker and Martha Schwartz
Design Team: Martha Schwartz, Ken Smith, David Meyer
Awards: American Society of Landscape Architects Honor Award, 1991

City and Nature Master Garden
Location: Xi'an, China
Client: Xi'an International Horticultural Exposition Organizing Committee
Year: 2011
Status: Completed
Design Team: Martha Schwartz, Donald Sharp, Allison Dailey, Cristabel Lee
Photographer: Gen Wang

City of Tomorrow Bo01
Location: Malmö, Sweden
Client: European Housing Expo
Year: 2001
Status: Completed
Design Team: Martha Schwartz, France Cormier

| 参与人员名单 | Credits | 344 |

戴维斯住宅
地点： 美国得克萨斯州埃尔帕索
客户： 萨姆和安妮·戴维斯
年份： 1996
状态： 已完成
设计团队： 玛莎·施瓦茨、迈克尔·布利尔、莎拉·费奇德、凯文·康格、保拉·梅杰灵克

德拉诺酒店
地点： 美国佛罗里达州迈阿密海滩
客户： 摩根酒店集团
年代： 1995
状态： 已完成
设计团队： 玛莎·施瓦茨、保拉·梅杰灵克、凯文·康格、莎拉·费奇德、克里斯·麦克法兰、劳拉·路特利奇、玛丽亚·贝尔阿尔塔

迪肯森住宅
地点： 美国新墨西哥州圣达菲
客户： 南希·迪肯森
年份： 1991
状态： 已完成
景观设计： 玛莎·施瓦茨工作室、肯·史密斯工作室、大卫·迈耶工作室
设计团队： 玛莎·施瓦茨、大卫·迈耶、莎拉·费奇德、肯·史密斯
摄影： 查尔斯·曼恩

迪斯尼乐园东入口广场
地点： 美国加利福尼亚州阿纳海姆
客户： 迪斯尼乐园
年份： 1998
状态： 已完成
设计团队： 玛莎·施瓦茨、唐纳德·夏普、保拉·梅杰灵克、李泰尔·费边、莎丽·韦斯曼、特里西娅·贝尔斯、伊夫林·伯加伊拉、斯科特·卡门、肖娜·吉利斯-史密斯、珍妮弗·布鲁克、拉斐尔·尤斯代威克、米歇尔·朗之万、韦斯·迈克尔斯、梅兰妮·米瑙特
摄影： 艾伦·沃德

多哈滨海大道设计竞赛
地点： 卡塔尔多哈
客户： 卡塔尔国家艺术与文化遗产委员会
年份： 2003
状态： 竞赛第二阶段
设计团队： 玛莎·施瓦茨、马库斯·詹斯奇、达伦·西尔斯、克劳迪娅·哈勒尔、唐纳德·布斯、伊莎贝尔、蔡姆拜尔、拉姆齐·巴达维、诺拉·李伯屯、路皮塔·贝兰加、克里斯蒂安·魏耶、周红
建筑设计： 詹斯奇-劳克斯建筑设计公司、马西建筑设计公司

交易广场
地点： 英国曼彻斯特
客户： 曼彻斯特市议会
年份： 1999
状态： 已完成
设计团队： 玛莎·施瓦茨、肖娜·吉利斯-史密斯、唐纳德·夏普、保拉·梅杰灵克、李泰尔·兹木克、费边、特里西娅·贝尔斯、韦斯·迈克尔斯、伊夫林·伯加伊拉、斯科特·卡门、拉斐尔·尤斯代威克、弗朗西斯卡·莱瓦吉、迈克尔·瓦瑟、詹姆斯·劳德、迈克尔·布利尔、罗德里克·怀利

Davis Residence
Location: El Paso, TX, USA
Client: Sam and Anne Davis
Year: 1996
Status: Completed
Design Team: Martha Schwartz, Michael Blier, Sara Fairchild, Kevin Conger, Paula Meijerink

Delano Hotel
Location: Miami Beach, FL, USA
Client: Morgan Hotel Group
Year: 1995
Status: Completed
Design Team: Martha Schwartz, Paula Meijerink, Kevin Conger, Sara Fairchild, Chris Macfarlane, Laura Rutledge, Maria Bellalta

Dickenson Residence
Location: Santa Fe, NM, USA
Client: Nancy Dickenson
Year: 1991
Status: Completed
Landscape Architect: The Office of Martha Schwartz, Ken Smith, David Meyer
Design Team: Martha Schwartz, David Meyer, Sara Fairchild, Ken Smith
Photographer: Charles Mann

Disneyland East Esplanade
Location: Anaheim, CA, USA
Client: Disneyland
Year: 1998
Status: Completed
Design Team: Martha Schwartz, Donald Sharp, Paula Meijerink, Lital Fabian, Sari Weissman, Tricia Bales, Evelyn Bergaila, Scott Carmen, Shauna Gilles-Smith, Jennifer Brooke, Rafael Justewicz, Michel Langevin, Wes Michaels, Melanie Mignault
Photographer: Alan Ward

Doha Corniche Competition
Location: Doha, Qatar
Client: The State of Qatar, National Council for Art and Heritage
Year: 2003
Status: 2. prize
Design Team: Martha Schwartz, Markus Jatsch, Darren Sears, Claudia Harari, Donald Booth, Isabel Zempel, Ramsey Badawi, Nora Libertun, Lupita Berlanga, Christian Weier, Hong Zhou
Architect: Jatsch Laux Architects, Massie Architecture

Exchange Square
Location: Manchester, UK
Client: Manchester Millenium
Year: 1999
Status: Completed
Design Team: Martha Schwartz, Shauna Gillies-Smith, Donald Sharp, Paula Meijerink, Lital Szmuk Fabian, Tricia Bales, Wes Michaels, Evelyn Bergaila, Scott Carmen, Rafael Justewicz, Francesca Levaggi, Michael Wasser, James Lord, Michael Blier, Roderick Wyllie

参与人员名单 | Credits

Fengming Mountain Park
Location: Chongqing, China
Client: Vanke
Year: 2013
Status: Completed
Design Team: Martha Schwartz, Nigel Koch, Jasmine Ong, Cristabel Lee, Aigars Lauzis, Ignacio Lopez-Buson, Ceylan Belek-Ombregt, Markus Jatsch, Francesca Neus Frontera, Gilles de Wever, Alicia Hidalgo, Don Sharp
Associate Landscape Architect: LaCime International
Photographer: Terrence Zhang

Fryston Village Green
Location: Castleford, England
Client: English Partnership
Year: 2004
Status: Completed
Design Team: Martha Schwartz, Matt Fougerat, John Pegg, Paula Craft, Sebastian Koepf, Steve Tycz, Trevor Lee, Angela Lim, Han Song Lee, Humberto Panti Garza, Isabel Zempel, Christian Bender, Christian Weier, David Mendelson, Daniel Booth, Meredith Redford, Sam Fleischmann

Garden Ornaments
Location: Dörentrup, Germany
Client: Bielefeld Art Museum
Year: 2001
Status: Completed
Design Team: Martha Schwartz, Isabel Zempel

Gifu Kitagata Gardens
Location: Gifu, Japan
Client: Gifu Prefectural Government
Year: 2000
Status: Completed
Design Team: Martha Schwartz, Paula Meijerink, Shauna Gillies-Smith, Michal Biller, Chris MacFarlane, Kak Martin, Don Sharp

Grand Canal Square
Location: Dublin, Ireland
Client: Dublin Docklands Development Corporation
Year: 2007
Status: Completed
Design Team: Martha Schwartz, Shauna Gillies-Smith, Donald Sharp, Friederike Huth, John Pegg, Paula Craft, Jessica Canfield, Christian Weier, Jay Rohrer, Sue Bailey, Isabel Zempel, Amit Arya, Christian Janssen, Courtney Pope, Heather Ring, James Cogliano, James Vincent, Joe Ficociello, Laura Knosp, Maria Bellata, Matt Fougerat, Rebecca Verner, Thomas Oles
Associate Landscape Architect: Tiros Resources
Lighting Designer: Speirs and Major Associates
Photographer: Tim Crocker, Tim Richardson

HUD Plaza
Location: Washington, D.C., USA
Client: U.S. General Services Administration
Year: 1996
Status: Completed
Design Team: Martha Schwartz, Evelyn Bergaila, Paula Mrijerink, Chris Macfarlane, Michael Blier, Kevin Conger, Dsts Fairchild, Scott Wunderle, Kaki Martin, David Bartsch, Rick Casteel
Awards: ASLA Merit Award 2000

参与人员名单 | Credits

雅各布·贾维茨广场
地点：美国纽约州纽约
客户：美国总务管理局
年份：1997
状态：已完成
设计团队：玛莎·施瓦茨、劳拉·路特利奇、玛丽亚·贝尔阿尔塔、克里斯·麦克法兰、迈克尔·布利尔、列诺·犹伊。
奖项：菲利普·N. 温斯洛景观设计奖，1997；美国风景园林师协会荣誉奖，1997；辛德勒国家设计奖，1998

金县监狱广场
地点：美国华盛顿州西雅图
客户：金县艺术委员会
年份：1987
状态：已完成
景观设计：彼得·沃克工作室、玛莎·施瓦茨工作室
设计团队：玛莎·施瓦茨、肯·史密斯、马丁·普瓦里埃、布拉德利·伯克
摄影：詹姆斯·发宁

林肯路购物中心
地点：美国佛罗里达州迈阿密海滩
客户：迈阿密市海滩
年份：1987
状态：已完成
设计团队：玛莎·施瓦茨、迈克尔·布利尔、保拉·梅杰灵克、克里斯·麦克法兰
建筑设计：汤普森·伍德建筑设计事务所、卡洛斯·萨帕塔设计事务所

露露岛
地点：阿联酋阿布扎比
客户：搜罗富房地产公司
年份：2006
状态：已完成
设计团队：玛莎·施瓦茨、唐纳德·夏普、莱斯利·李、玛丽亚·贝尔阿尔塔、肖娜·吉利斯-史密斯
建筑设计：阿魁国际设计公司

梅萨艺术中心
地点：美国亚利桑那州梅萨
客户：梅萨市
年份：2005
状态：已完成
设计团队：玛莎·施瓦茨、肖娜·吉利斯-史密斯、唐纳德·夏普、弗兰西·科米尔、罗伊·费边、克里斯蒂娜·帕特森、克里斯塔尔·英格兰、莎丽·韦瓦斯曼、迈克尔·格吕克、尼考勒·根茨勒、李泰尔·费边、迈克尔·基凯利、内特·特雷弗汉、韦斯·迈克尔斯、苏珊·奥尼拉斯、帕特里夏·贝斯、保拉·梅杰灵克
景观设计合作方：设计工作室
建筑设计：博拉建筑设计事务所
喷泉设计：丹·欧瑟（水利建筑公司）

滨海线性公园
地点：美国加利福尼亚州圣地亚哥
客户：中心城区开发公司
年份：1988
状态：已完成
景观设计：彼得·沃克工作室、玛莎·施瓦茨工作室
设计团队：彼得·沃克、玛莎·施瓦茨
奖项：美国风景园林师协会优秀设计奖，1991

Jacob Javits Plaza
Location: New York, NY, USA
Client: U.S. General Services Administration
Year: 1997
Status: Completed
Design Team: Martha Schwartz, Laura Rutledge, Maria Bellalta, Chris MacFalane, Michael Blier, Leo Jew
Awards: Phillip N. Winslow Landscape Design Award 1997, ASLA Honor Award 1997, GSA Design Award National Design Citation 1998

King County Jailhouse Plaza
Location: Seattle, WA, USA
Client: King County Arts Commission
Year: 1987
Status: Completed
Landscape Architect: The Office of Peter Walker and Martha Schwartz
Design Team: Martha Schwartz, Ken Smith, Martin Poirier, Bradley Burke
Photography: James Faning

Lincoln Road Mall
Location: Miami Beach, FL, USA
Client: City of Miami Beach
Year: 1987
Status: Completed
Design Team: Martha Schwartz, Michael Blier, Paula Meijerink, Chris McFarlane
Architect: Thompson and Wood, Architects & Carlos Zapata Design Studio

Lulu Island
Location: Abu Dhabi, UAE
Client: Sorough Real Estate
Year: 2006
Status: Completed
Design Team: Martha Schwartz, Donald Sharp, Leslie Li, Maria Bellalta, Shauna Gillies-Smith
Architect: Arquitectonica International

Mesa Arts Center
Location: Mesa, AZ, USA
Client: City of Mesa
Year: 2005
Status: Completed
Design Team: Martha Schwartz, Shauna Gillies-Smith, Donald Sharp, France Cormier, Roy Fabian, Kristina Patterson, Krystal England, Sari Weissman, Michael Glueck, Nicole Gaenzler, Lital Szmuk Fabian, Michael Kilkelly, Nate Trevethan, Wes Michaels, Susan Ornelas, Patricia Bales, Paula Meijerink
Associate Landscape Architect: Design Workshop
Architect: Boora Architects
Fountain Designer: Dan Euser (Waterarchitecture)

Marina Linear Park
Location: San Diego, CA, USA
Client: Center City Development Corporation
Year: 1988
Status: Completed
Landscape Architect: The Office of Peter Walker and Martha Schwartz
Design Team: Peter Walker, Martha Schwartz
Award: American Society of Landscape Architects Merit Award 1991

| 参与人员名单 | Credits |

麦克劳德尾矿
地点：加拿大杰拉尔顿
客户：巴里克黄金公司
年份：1998
状态：已完成
设计团队：玛莎·施瓦茨、李泰尔·兹木克·费边、詹姆斯·劳德、特里西娅·贝尔斯、肖娜·吉利斯-史密斯
合作方：库克工程公司（保罗·布鲁格）

迈阿密国际机场隔音墙
地点：美国佛罗里达州迈阿密
客户：戴德地铁公共艺术公司
年份：1996
状态：已完成
设计团队：玛莎·施瓦茨、凯文·康格、莎拉·费奇德、克里斯·麦克法兰、劳拉·路特利奇、玛丽亚·贝尔阿尔塔、列诺·犹伊
获奖：戴德地铁公共艺术设计奖

明尼阿波利斯联邦法院广场
地点：美国明尼苏达州明尼阿波利斯
客户：美国总务管理局
年份：1996
状态：已完成
设计团队：玛莎·施瓦茨、保拉·梅杰灵克、克里斯·麦克法兰、劳拉·路特利奇、玛丽亚·贝尔阿尔塔、列诺·犹伊
获奖：辛德勒国家设计奖，1996年；美国风景园林学会优秀设计奖，1999年；美国国家艺术基金会联邦设计成就奖，2000

蒙特拉中央公园
地点：奥地利维也纳
客户：奥地利PORR隧道建设公司
年份：2007
状态：已完成
设计团队：玛莎·施瓦茨、伊莎贝尔·蔡姆拜尔、尼考勒·根茨勒、弗朗西·科米尔、保拉·梅杰灵克、莎丽·韦斯曼、弗里德里克·胡特、诺拉·李伯屯
景观设计合作方：3:0景观设计公司
建筑设计：阿尔伯特·维莫建筑设计公司

莫斯科儿童休闲街区
地点：俄罗斯莫斯科
客户：斯特列尔卡公司
年份：2015
状态：已完成
设计团队：玛莎·施瓦茨、马库斯·詹斯奇、迭戈·帕切科·冈萨雷斯、杰兰·贝莱克·奥姆雷特

海军码头公园设计竞赛
地点：美国伊利诺伊州芝加哥
客户：海军码头公园
年份：2012
设计团队：玛莎·施瓦茨、马库斯·詹斯奇、马修·盖驰、爱加斯·劳兹斯
建筑城市规划：戴维斯·布罗迪·邦德建筑设计事务所／Aedas建筑设计事务所、马歇尔·布朗设计室、所罗门·科德韦尔·布恩泽建筑设计事务所

McLeod Tailings
Location: Geraldton, ON, Canada
Year: 1998
Status: Completed
Design Team: Martha Schwartz, Lital Szmuk Fabian, James Lord, Tricia Bales, Shauna Gillies-Smith
Collaborator: Cook Engineering (Paul Brugger)

Miami International Airport Sound Wall
Location: Miami, FL, USA
Client: Metro-Dade Art in Public Places
Year: 1996
Status: Completed
Design Team: Martha Schwartz, Kevin Conger, Sara Fairchild, Chris Macfarlane, Laura Rutledge, Maria Bellalta, Leo Jew
Award: Metro-Dade Art in Public Places Award

Minneapolis Courthouse Plaza
Location: Minneapolis, MN, USA
Client: U.S. General Services Administration
Year: 1996
Status: Completed
Design Team: Martha Schwartz, Paula Meijerink, Chris McFarlane, Laura Rutledge, Maria Ballata, Leo Jew
Awards: GSA Design Award National Design Citation 1996, ASLA Merit Award 1999, NEA Federal Design Achievement Award 2000

Monte Laa Central Park
Location: Vienna, Austria
Client: Porr
Year: 2007
Status: Completed
Design Team: Martha Schwartz, Isabel Zempel, Nicole Gaenzler, France Cormier, Paula Meijerink, Sari Weissman, Friederike Huth, Nora Libertun
Associate Landscape Architect: 3:0 Landschaftsarchitektur
Architect: Atelier Albert Wimmer

Moscow Children's Route
Location: Moscow, Russia
Client: Strelka KB
Year: 2015
Status: Completed
Design Team: Martha Schwartz, Markus Jatsch, Diego Pacheco Gonzalez, Ceylan Belek Ombregt

Navy Pier Competition
Location: Chicago, IL, USA
Client: Navy Pier
Year: 2012
Design Team: Martha Schwartz, Markus Jatsch, Matthew Getch, Aigars Lauzis
Architects, Urban Planners: Davis Brody Bond / Aedas, Marshal Brown Projects, Solomon Cordwel Buenz

		348

参与人员名单 　　　　　　　　　　　　　　　　Credits

联排住宅
地点：日本福冈
客户：福冈吉首
年份：1991
状态：已完成
景观设计：玛莎·施瓦茨工作室、肯·史密斯工作室、大卫·迈耶工作室
设计团队：玛莎·施瓦茨、肯·史密斯、大卫·迈耶、凯瑟琳·德灵豪斯、斯科特·萨默、格尔达·亚历山大、保拉·梅杰灵克、迈克尔·布利尔、克里斯·麦克法兰、凯文·康格
摄影：理查德·巴恩斯

北钓鱼台开发项目
地点：中国北京
客户：北京长实东方地产公司
年份：2014
状态：已完成
设计团队：玛莎·施瓦茨、马修·盖驰、马库斯·詹斯奇、丹·雷亚、崔波、安德诺尼基·施特朗吉罗、西莱纳·普萨里多、尹珠利、吉列斯·得·威弗、Kyung Eui Park、李奇迪、伊迪丝·卡茨
摄影：齐飞

纳提克努韦勒屋顶花园
地点：美国马萨诸塞州纳提克
客户：GGP地产集团
年份：2008
状态：已完成
设计团队：玛莎·施瓦茨、肖娜·吉利斯-史密斯
摄影：Chuck Choi
奖项：波士顿风景园林学会优秀设计奖，2012

保罗林克庭院
地点：德国柏林
客户：里尔房地产公司
年份：2000
状态：已完成
设计团队：玛莎·施瓦茨、保拉·梅杰灵克、帕特里西亚·莱瓦吉、韦斯·迈克尔斯、肖娜·吉利斯-史密斯、迈克尔·瓦瑟、詹姆斯·劳德
景观设计合作方：基弗工作室

共和广场
地点：法国巴黎
客户：巴黎路政署
年份：2013
状态：已完成
设计团队：玛莎·施瓦茨、马修·盖驰
景观设计合作方：阿里尔公司
建筑设计：TVK建筑设计事务所
摄影：克莱门特·纪尧姆

普鲁伊特市规划
地点：印度尼西亚雅加达
客户：阿贡公司
状态：规划中
设计团队：玛莎·施瓦茨、马库斯·詹斯奇、马修·盖驰、埃克·塞尔比、艾丽西亚·伊达尔戈·洛佩兹
建筑设计：SOM建筑设计事务所

Nexus Housing
Location: Fukuoka, Japan
Client: Fukuoka Jisho
Year: 1991
Status: Completed
Landscape Architect: The Office of Martha Schwartz, Ken Smith, David Meyer
Design Team: Martha Schwartz, Ken Smith, David Meyer, Kathryn Drinkhouse, Scott Summers, Verda Alexander, Paula Meijerink, Michael Blier, Chris MacFarlane, Kevin Conger
Photography: Richard Barnes

North Diaoyutai Development
Location: Beijing, China
Client: Beijing Changshi Oriental Land
Year: 2014
Status: Completed
Design Team: Martha Schwartz, Matthew Getch, Markus Jatsch, Dan Rea, Bo Cui, Androniki Strongioglou, Silena Ptsalidou, Yoon Joo Lee, Gilles de Wever, Kyung Eui Park, Qidi Li, Edith Katz
Photographer: Fei Qi

Nouvelle at Natick
Location: Natick, MA, USA
Client: General Growth Properties
Year: 2008
Status: Completed
Design Team: Martha Schwartz, Shauna Gillies-Smith
Photographer: Chuck Choi
Awards: BSLA Award 2012

Paul-Lincke-Höfe
Location: Berlin, Germany
Client: Realprojekt Bau– und Boden
Year: 2000
Status: Completed
Design Team: Martha Schwartz, Paula Meijerink, Patricia Levaggi, Wes Michael, Shauna Gillies-Smith, Michael Wasser, James Lord
Associate Landscape Architect: Buro Kiefer

Place de la République
Location: France, Paris
Client: City of Paris, Highways Department
Year: 2013
Status: Completed
Design Team: Martha Schwartz, Matthew Getch
Associate Landscape Architect: AREAL
Architects: TVK Architectes Urbanistes
Photographer: Clement Guillaume

Pluit City
Location: Jakarta, Indonesia
Client: Agung Podomoro Land
Status: In planning
Design Team: Martha Schwartz, Markus Jatsch, Matthew Getch, Eike Selby, Alicia Hidalgo Lopez
Architect: SOM

| 参与人员名单 | Credits |

输电线景观
地点：德国盖尔森基兴
客户：国际建筑博览会埃姆歇公园
年份：1998
状态：已完成
设计团队：玛莎·施瓦茨、马库斯·詹斯奇

里约购物中心
地点：美国佐治亚州亚特兰大
客户：阿克曼公司
年份：1989
状态：已完成
景观设计：彼得·沃克工作室、玛莎·施瓦茨工作室
设计团队：玛莎·施瓦茨、肯·史密斯、大卫·迈耶、马丁·普瓦里埃、道格·芬德利、大卫·沃克
奖项：美国风景园林学会优秀设计奖，1989

圣玛丽教堂公园
地点：英国伦敦
客户：伦敦南华克区
年份：2008
状态：已完成
设计团队：玛莎·施瓦茨、埃达·奥斯特塔格、弗里德里克·胡特、苏珊·贝利、Annghi Tran、德博拉·纳甘、丹尼尔·雷亚、克里斯·贝利、Kwong Hang Wong、克里斯蒂安·魏耶、克劳迪娅·施拖尔特、雅科·奈尔、约翰·佩格、朱利安·博勒特、劳里·普雷布尔、莱顿·佩斯、马库斯·希尔兹、马特·富捷拉特、奈杰尔·索恩、保拉·克拉夫特、西蒙娜·马什

索沃广场
项目位置：阿联酋阿布扎比
客户：穆巴达拉开发公司
年份：2012
状态：已完成
设计团队：玛莎·施瓦茨、马修·盖驰、纳吉尔·考赫、彼得·皮特、克丽丝特·李
建筑设计：格驰及其合伙人建筑设计事务所/晋思建筑设计事务所
摄影：邓肯·查德

拼贴花园
地点：美国马萨诸塞州剑桥
客户：怀特黑德生物医学研究所
年份：1986
状态：已完成
景观设计：彼得·沃克工作室、玛莎·施瓦茨工作室
设计团队：玛莎·施瓦茨、布拉德利·伯克
摄影：艾伦·沃德

斯波莱托艺术节景观
地点：美国南卡罗来纳州查尔斯顿
客户：美国斯波莱托艺术节管理委员会
年份：1997
状态：已完成
设计团队：玛莎·施瓦茨、丽莎·戴尔普拉、李泰尔、费边、伊夫林·伯加伊拉、韦斯·迈克尔斯、卡奇·马丁

Power Lines
Location: Gelsenkirchen, Germany
Client: International Building Exhibition Emscher Park
Year: 1998
Status: Completed
Design Team: Martha Schwartz, Markus Jatsch

Rio Shopping Center
Location: Atlanta, GA, USA
Client: Ackerman and Company
Year: 1989
Status: Completed
Landscape Architect: The Office of Peter Walker and Martha Schwartz
Design Team: Martha Schwartz, Ken Smith, David Meyer, Martin Poirier, Doug Findlay, David Walker
Award: ASLA Merit Award 1989

Saint Mary's Churchyard Park
Location: London, UK
Client: London Borough of Southwark Council
Year: 2008
Status: Completed
Design Team: Martha Schwartz, Edda Ostertag, Friederike Huth, Susan Bailey, Annghi Tran, Deborah Nagan, Daniel Rea, Chris Bailey, Kwong Hang Wong, Christian Weier, Claudia Stolte, Jaco Nel, John Pegg, Julian Bolleter, Laurie Preble, Leighton Pace, Marcus Shields, Matt Fougerat, Nigel Thorne, Paula Craft, Simone Marsh

Sowwah Square
Location: Abu Dhabi, UAE
Client: Mubadala Development
Year: 2012
Status: Completed
Design Team: Martha Schwartz, Matthew Getch, Nigel Koch, Peter Piet, Cristabel Lee
Architect: Goettsch Partners and Gensler
Photographer: Duncan Chard

Splice Garden
Location: Cambridge, MA, USA
Client: Whitehead Institute for Biomedical Research
Year: 1986
Status: Completed
Landscape Architect: The Office of Peter Walker and Martha Schwartz
Design Team: Martha Schwartz, Bradley Burke
Photographer: Alan Ward

Spoleto Festival
Location: Charleston, SC, USA
Client: Spoleto Festival USA
Year: 1997
Status: Completed
Design Team: Martha Schwartz, Lisa Delplace, Lital Fabian, Evelyn Bergaila, Wes Michaels, Kaki Martin

| 参与人员名单 | Credits | 350 |

卢森堡南部医院景观
地点：卢森堡阿尔泽特/埃施
客户：爱米勒·玛里奇中心医院
年份：2015
状态：竞赛第一阶段（规划中）
设计团队：玛莎·施瓦茨、马库斯·詹斯奇、Yoon Joo Lee
建筑设计：阿尔伯特·维莫建筑设计公司

瑞士再保险总部大楼
地点：德国慕尼黑
客户：巴伐利亚
年份：2002
状态：已完成
设计团队：玛莎·施瓦茨、保拉·梅杰灵克、特里西娅·贝尔斯、韦斯·迈克尔斯、米歇尔·朗之万、梅兰妮·米瑙特、肖娜·吉利斯-史密斯、迈克尔·格吕克、李泰尔·费边、克里斯塔尔·英格兰、尼考勒·根茨勒
景观设计合作方：彼得·克鲁斯卡
建筑设计：BRT建筑设计公司
摄影：Myrzik、Jarisch

万科中心
地点：中国深圳
客户：深圳万科房地产开发公司
年份：2013
状态：已完成
设计团队：玛莎·施瓦茨、克里斯·王、唐纳德·夏普、纳吉尔·考赫、Soojung Rhee
建筑设计：斯蒂芬·霍尔
摄影：泰伦斯·张

维也纳北方医院景观
地点：奥地利维也纳
客户：维也纳疾病预防中心
年份：2008
状态：竞赛第一阶段（规划中）
设计团队：玛莎·施瓦茨、马修·盖驰
景观设计合作方：3：0景观设计公司
建筑设计：阿尔伯特·维莫建筑设计公司

约克维尔公园
地点：加拿大多伦多
客户：多伦多市公园休闲局
年份：1995
状态：已完成
景观设计：玛莎·施瓦茨工作室、肯·史密斯工作室、大卫·迈耶工作室
设计团队：玛莎·施瓦茨、肯·史密斯、大卫·迈耶
摄影：彼得·毛斯
奖项：美国风景园林学会会长奖，1994；美国风景园林学会优秀设计会长奖，1996；多伦多市城市设计奖，1997；美国风景园林学会地标奖，2012

温斯洛农场保护景观
地点：美国新泽西州哈蒙顿
客户：汉克·麦克尼尔
年份：1996
状态：已完成
设计团队：玛莎·施瓦茨、凯瑟琳·德灵豪斯、迈克尔·布利尔、凯文·康格、保拉·梅杰灵克、伊夫林·伯加伊拉、李泰尔·费边、梅兰妮、米瑙特、米歇尔·朗之万

Sudspidol Luxemburg
Location: Esch/Alzette, Luxemburg
Client: Centre Hospitalier Emile Mayrisch
Year: 2015
Status: Competition, 1. prize (in planning)
Design Team: Martha Schwartz, Markus Jatsch, Yoon Joo Lee
Architect: Atelier Albert Wimmer

Swiss Re Headquarters
Location: Munich, Germany
Client: Bayerische Rück
Year: 2002
Status: Completed
Design Team: Martha Schwartz, Paula Meijerink, Tricia Bales, Wes Michaels, Michel Langevin, Melanie Mignault, Shauna Gillies-Smith, Michel Glueck, Lital Fabian, Krystal England, Nicole Gaenzler
Associated Landscape Architect: Peter Kluska
Architects: BRT Architekten
Photographer: Myrzik and Jarisch

Vanke Center
Location: Shenzhen, China
Client: Shenzhen Vanke Real Estate
Year: 2013
Status: Completed
Design Team: Martha Schwartz, Chris Wong, Don Sharp, Nigel Koch, Soojung Rhee
Architect: Steven Holl
Photographer: Terrence Zhang

Vienna North Hospital
Location: Vienna, Austria
Client: Wiener Krankenanstaltenverbund
Year: 2008
Status: Competition, 1. prize (in planning)
Design Team: Martha Schwartz, Matthew Getch
Associate Landscape Architect: 3:0 Landschaftsarchitekten
Architect: Atelier Albert Wimmer

Village of Yorkville Park
Location: Toronto, Canada
Client: City of Toronto, Department of Parks and Recreation
Year: 1995
Status: Completed
Landscape Architect: The Office of Martha Schwartz, Ken Smith, David Meyer
Design Team: Martha Schwartz, Ken Smith, David Meyer
Photographer: Peter Maus
Awards: ASLA President's Award 1994, ASLA President's Award of Excellence 1996, City of Toronto Urban Design Award 1997, ASLA Landmark Award 2012

Winslow Farm Conservancy
Location: Hammonton, NJ, USA
Client: Hank McNeil
Year: 1996
Status: Completed
Design Team: Martha Schwartz, Kathryn Drinkhouse, Michael Blier, Kevin Conger, Paula Meijerink, Evelyn Bergeila, Lital Szmuk-Fabian, Melanie Mignault, Michel Langevin

| 参与人员名单 | Credits | 351 |

龙山国际商务区
地点：韩国首尔
客户：龙山开发公司、理想金融投资公司
年份：2012
设计团队：玛莎·施瓦茨、马库斯·詹斯奇、爱加斯·劳兹斯、克里斯·王、Soojung Rhee
建筑设计：丹尼尔·里伯斯金工作室（总体规划，R1，B2-1）；BIG（R4b）；MVRDV（R4a）；阿德里安·史密斯+戈登·吉尔建筑设计事务所（B4）

Yongsan International Business District
Location: Seoul, South Korea
Client: Yongsan Development, Dreamhub Project Financial Investment
Year: 2012
Design Team: Martha Schwartz, Markus Jatsch, Aigars Lauzis, Chris Wong, Soojung Rhee
Architects: Studio Daniel Libeskind, BIG, MVRDV, Adrian Smith + Gordon Gill Architecture

图书在版编目（CIP）数据

景观艺术与城市设计：玛莎·施瓦茨及合伙人设计事务所作品集 /（德）詹斯奇（Jatsch, M.）主编；杨至德译. -- 南京：江苏凤凰科学技术出版社，2016.4
 ISBN 978-7-5537-4176-5

Ⅰ.①景 Ⅱ.①詹 ②杨 Ⅲ.①景观设计－作品集－德国－现代 Ⅳ.①TU-881.516

中国版本图书馆CIP数据核字(2015)第039862号

景观艺术与城市设计
玛莎·施瓦茨及合伙人设计事务所作品集

主　　　编	[德] 马库斯·詹斯奇（Markus Jatsch）
译　　　者	杨至德
项 目 策 划	凤凰空间/高雅婷
责 任 编 辑	刘屹立
特 约 编 辑	林　溪
出 版 发 行	凤凰出版传媒股份有限公司
	江苏凤凰科学技术出版社
出版社地址	南京市湖南路1号A楼，邮编：210009
出版社网址	http://www.pspress.cn
总　经　销	天津凤凰空间文化传媒有限公司
总经销网址	http://www.ifengspace.cn
经　　　销	全国新华书店
印　　　刷	北京盛通印刷股份有限公司
开　　　本	965 mm×1 270 mm　1/16
印　　　张	22
字　　　数	422 000
版　　　次	2016年4月第1版
印　　　次	2016年4月第1次印刷
标 准 书 号	ISBN 978-7-5537-4176-5
定　　　价	298.00元（USD 45.00）（精）

图书如有印装质量问题，可随时向销售部调换（电话：022-87893668）。